The Beauty of Doing Mathematics

Serge Lang

The Beauty of
Doing Mathematics
Three Public Dialogues

With 91 Illustrations

Springer-Verlag
New York Berlin Heidelberg Tokyo

Serge Lang
Department of Mathematics
Yale University
New Haven, CT 06520
U.S.A.

AMS Classifications: 00-AO5, 00-AO6

Cover photo courtesy of Carol MacPherson.

Library of Congress Cataloging in Publication Data
Lang, Serge
  The beauty of doing mathematics.
  "Originally they were published in the Revue du
Palais de la Découverte [in French]"—Pref.
  Contents: What does a pure mathematician do and why?—
prime numbers—To do mathematics, a lively activity—
diophantine equations—Great problems of geometry and
space.
  Bibliography: p.
  1. Mathematics—Addresses, essays, lectures. I. Title.
QA7.L28   1985   510   85-13838

Translation of: *Serge Lang, Fait des Maths en Public*, Belin, 1984.

Typeset by House of Equations, Inc., Newton, New Jersey.
Printed and bound by Halliday Lithograph, West Hanover, Massachusetts.
Printed in the United States of America.

9 8 7 6 5 4 3 2 1

ISBN 0-387-96149-6 Springer-Verlag New York Berlin Heidelberg Tokyo
ISBN 3-540-96149-6 Springer-Verlag Berlin Heidelberg New York Tokyo

Preface

If someone told you that mathematics is quite beautiful, you might be surprised. But you should know that some people do mathematics all their lives, and create mathematics, just as a composer creates music. Usually, every time a mathematician solves a problem, this gives rise to many others, new and just as beautiful as the one which was solved. Of course, often these problems are quite difficult, and as in other disciplines can be understood only by those who have studied the subject with some depth, and know the subject well.

In 1981, Jean Brette, who is responsible for the Mathematics Section of the Palais de la Découverte (Science Museum) in Paris, invited me to give a conference at the Palais. I had never given such a conference before, to a non-mathematical public. Here was a challenge: could I communicate to such a Saturday afternoon audience what it means to do mathematics, and why one does mathematics? By "mathematics" I mean pure mathematics. This doesn't mean that pure math is better than other types of math, but I and a number of others do pure mathematics, and it's about them that I am now concerned.

Math has a bad reputation, stemming from the most elementary levels. The word is in fact used in many different contexts. First, I had to explain briefly these possible contexts, and the one with which I wanted to deal. Many people raised questions on a variety of topics: pure math, applied math, concrete versus abstract, the teaching of mathematics, and others which gave rise to a lively dialogue. But mostly, I wanted to show what pure mathematics is by examples: by doing mathematics with the people in the audience. And not artificial or superficial mathematics, but real mathematics, recognized as such by real mathematicians who do research in mathematics. So I had to find some topics which on the one hand were accessible to the Saturday afternoon public, who doesn't want to get bored or snowed, but who wants to learn without having any particular background in mathematics.

On the other hand, the subjects which I chose had to come from deep mathematics, they had to show great unsolved problems, which mathematicians actually work on. To discover new mathematics is the essence of what mathematicians do. The more you know, the more you realize how much you don't know. I could not cheat. I had to do real mathematics.

That's what I did the first time, in 1981. It worked so well that I came back twice after that, in 1982 and 1983, each time choosing a different topic. The first two are rather algebraic: prime numbers and diophantine equations; while the third is geometric: great problems of geometry and space. The third time was a true marathon, during which a hundred persons stayed more than three and half hours! I had never expected that one could achieve such a result. I was extremely touched by the audience reaction, all three times, but especially this last time.

I emphasize that the audience was not principally composed of mathematicians. I asked the few of them who were present not to intervene. The general audience was very diverse, ranging from young high school (even some grammar school) students, to retired people, housewives, engineers, or just plain curious people. They participated actively and themselves raised interesting questions.

The three talks are independent of each other. So you don't have to read them in any particular order. Each one forms a single whole, which you can appreciate independently of the others. While reading, if you come across something which seems difficult or too rough, then don't be discouraged. Skip it, go on to the next paragraph, or next page, or next lecture. Even within the same talk you'll very likely find some things which again make sense or appeal to you, and which will appear easier. If you are still interested, you might come back later to those parts which bothered you. You will be surprised to see how often, after sleeping on it, something which appeared difficult suddenly seems easy.

A good part of the program of elementary or secondary schools is very arid and technical, and you may never have had the luck to see how mathematics can be beautiful. If you are a student, in high school or college, I hope you will find here something to complement the math courses which you have taken, or which have been imposed on you, whatever they were.

This book reproduces the three talks I gave in Paris. They were transcribed from tapes as faithfully as possible, so as to keep the lively style. Originally, they were published in the *Revue du Palais de la Découverte* (the journal of the Science Museum). I am very thankful to Brette for his encouragement, and for his cooperation in transcribing the tapes, as well as for the drawings and figures which are due to him. Other people thought the three talks should be published in a single volume. I thank the publishing house Belin, and in particular its Director Max Brossollet, for their enthusiasm which gave rise to the first publication in French, and I thank Springer-Verlag for the English version which is now appearing. Finally, I thank Carol MacPherson for the photograph on the cover. I also thank the Palais de la Découverte for its cooperation. Everyone contributed to make it possible to preserve, as far as possible, the spontaneity of the original conferences, and the framework in which they took place.

SERGE LANG

Contents

Who is Serge Lang?

Serge Lang was born in Paris in 1927. He went to school until the 10th grade in the suburbs of Paris, where he lived. Then he moved to the United States. He did two years of high school in California, then entered the California Institute of Technology (Caltech), from which he graduated in 1946. After a year and a half in the American army, he went to Princeton in the Philosophy Department where he spent a year. He then switched to mathematics, also at Princeton, and received his PhD in 1951. He taught at the university and spent a year at the Institute for Advanced Study, which is also in Princeton.

Then he got into more regular positions: Instructor at the University of Chicago, 1953–1955; Professor at Columbia University, 1955–1970. In between, he spent a year as a Fulbright scholar in Paris in 1958.

He left Columbia in 1970. He was Visiting Professor at Princeton in 1970–1971, and Harvard in 1971–1972. Since 1972 he has been a professor at Yale.

Besides math, he mostly likes music. During different periods of his life, he played the piano and the lute.

From 1966 to 1969, Serge Lang was politically and socially active, during a period when the United States faced numerous problems which affected the universities very deeply.

He has also been concerned with the problems of financing the universities, and of their intellectual freedom, threatened by political and bureaucratic interference. As he says, such problems are invariant under ism transformations: socialism, communism, capitalism, or any other ism in the ology.

However, his principal interest has always been for mathematics. He has published 28 books and more than 60 research articles. He received the Cole Prize in the U.S. and Prix Carriere in France.

What does a mathematician do and why?

Prime numbers

16 May 1981

Summary: *The conference started with why, for ten minutes. I do mathematics because I like it. We discussed briefly the distinction between pure mathematics and applied mathematics, which actually intermingle in such a way that it is impossible to define the boundary between one and the other precisely; and also the aesthetic side of mathematics. Then we did mathematics together. I started by defining prime numbers, and I recalled Euclid's proof that there are infinitely many. Then I defined twin primes, (3, 5), (5, 7), (11, 13), (17, 19), etc. which differ by 2. Is there an infinite number of those? No one knows, even though one conjectures that the answer is yes. I gave heuristic arguments describing the expected density of such primes. Why don't you try to prove it? The question is one of the big unsolved problems of mathematics.*

Almost every Saturday, from October to June, the Palais de la Découverte (Science Museum in Paris) traditionally welcomes and presents to the public eminent lecturers in all disciplines.

Thus we were honored to welcome Professor Serge Lang who is a world renowned mathematician, during a brief stay in Paris. He is the author of more than twenty seven mathematical books devoted both to teaching and research. When I invited Serge Lang, I knew him by reputation, and by some of his works. Therefore I had no worry on a purely mathematical plane.

I had only one apprehension, however: would he be a good lecturer? Would he know how to relate to a large non-mathematical public? As I expressed these thoughts to him shortly before the conference, sitting at a café, he told me that a good teacher is not only a specialist in his discipline, but also an actor, sensitive to the public's reactions. He also explained that he was very happy to have this new experience: talk and do mathematics with people who are not students, and show them what mathematics is by "doing mathematics" with them.

"And," he added laughing, "you will see what happens!" I saw! And the conference was a success! Of course, people were surprised to have to participate, and not only to listen; however, after a few minutes, Serge Lang's fiery spirit won them over, and the dialogue went on.

There remained the question of publication. Rather than a technical version, it seemed better to publish the whole conference and the questions that followed it, except for minor changes. Indeed, besides the actual mathematical content, it seemed to me useful to preserve, as far as was possible, the dynamic quality of this conference; the liveliness of the dialogue; and, why not, the actor's performance. Neither those present that day nor popularizers and pedagogues will complain about this.

So I submitted to him a first version, transcribed from the tapes of the lecture. Being very careful and concerned with precision, (which is also one of his personal characteristics), Serge Lang not only checked, but also retyped the whole text. In so doing, he had to familiarize himself with our computer terminals, the only ones which had an American keyboard! So he came every day for more than a week, conciliatory for certain changes, uncompromising as far as the style was concerned, choosing the most appropriate words, and adding a few items, especially on the Riemann Hypothesis, as well as a short bibliography on the topics which were discussed.

Jean Brette
Responsible for the Mathematics
Section of the Palais de la
Découverte

The conference

So, the conference, I think I'm going to talk about things in general for ten minutes, and after that, we'll try to do mathematics together. The talk will be as the title says: "What does a pure mathematician do and why?"

It's very difficult to explain "why" in a general way, and also what we do, in a general way. For example, "mathematics" is a word which is used for a lot of activities which don't have much to do with each other. I am sure that the word means very different things for different people. For instance, you, Madam [*Serge Lang points to a lady in the audience*], what does "mathematics" mean to you?

LADY. The abstraction of numbers, the manipulation of numbers.

SERGE LANG. In fact, one can do mathematics without using numbers at all; as in geometry, or spatial mathematics. It's true that to give you an example of mathematics, as I shall do a little later, I shall use numbers, but in a context which, I think, will be different from the one you are thinking about. And you, sir, what does it mean, "mathematics"?

GENTLEMAN. The manipulation of structures.

SERGE LANG. Yes, but which ones? There are lots of structures which are not mathematical. Mathematics is not just a question of structures. For example, when you do physics, you also manipulate certain structures. In fact, the word "mathematics" is used in many different contexts. You have mathematics as they are done in elementary or high school. You have computer mathematics, applied to problems of communications. If you are into physics or chemistry, you use mathematics to describe the empirical world. But what I want to talk about today is what I will call "pure mathematics", those which are done from a purely aesthetic point of view. To do mathematics like that is very different from studying the empirical world. It's different from describing or classifying the empirical world by means of mathematical models. An experimental scientist makes a choice among many possible models, to find those which fit the empirical world, the world of experiments, in trying to find a system for the world. There are lots of pure mathematics which are not used in studying the empirical world, and which are considered solely for their beauty. And this has been the case forever, for centuries, since there have been civilizations—Arabic, Hindu, whatever. The Greeks did mathematics for the beauty of it.[1]

It is true that some parts of mathematics have their source in the empirical world, but much mathematics is done independently of these sources. This point of view has been expressed by other mathematicians,

---

[1] Which does not exclude that they also did mathematics which had practical applications. Everyone agrees to include physics, chemistry, biology, under the general heading of "science". To decide whether "pure mathematics" as I have described them should also be placed under this heading is a question of terminology which I don't want to get into now.

and I want to read you something written by other mathematicians, for instance on the relation between doing mathematics as they relate to applied math.

Jacobi, who was a mathematician of the 19th century, wrote in a letter to Legendre:[2]

> I read with pleasure Mr. Poisson's report on my work, and think I can be very satisfied by it . . . but Mr. Poisson should not have reproduced a rather clumsy phrase by Mr. Fourier, who reproached Abel and me for not having preferred to work on heat flow. It is true that Mr. Fourier thought that the principal goal of mathematics was their public utility and their use in explaining natural phenomena. A philosopher like him should have known that the only goal of Science is the honor of the human spirit, and that as such, a question in number theory is worth a question concerning the system of the world.

In an article which appeared in the collection "Great Currents of Mathematical Thought", directed by F. Le Lionnais in 1948, Andre Weil (who is one of the great mathematicians of this century), quoted Jacobi in the following context:

> But if, like Panurge, we ask the oracle questions which are too indiscreet, then the oracle will answer as to Panurge: "Drink!" Advice which the mathematician is only too glad to follow, satisfied that he is to quench his thirst at the very sources of knowledge, satisfied that these sources always gush pure and abundant, while others must have recourse to the muddy paths of a sordid actuality. That if one reproaches him for his arrogant attitude, if one challenges him to engage himself in the actual world, if one asks why he persists on these high glaciers where none but others like him can follow him, he answers with Jacobi: "For the honor of the human spirit!"[3]

OK, that's literature. It's also a pompous style, which does not reflect accurately Jacobi's thoughts. To refer to others, who "must have recourse to the muddy paths of a sordid actuality" is not exactly the same thing as to say that "a question of number theory is worth a question concerning the system of the world". Weil, elsewhere, described in another way his

---

[2] No date, stamped 2 July 1830, *Collected Works of Jacobi*, Vol. 1, p. 454.

[3] The original is in French, and very literary French at that:

> Mais si, comme Panurge, nous posons a l'oracle des questions trop indiscretes, l'Oracle nous répondra comme à Panurge: Trinck! Conseil auquel le mathématicien obéit volontiers, satisfait qu'il est de croire étancher sa soif aux sources memes du savoir, satisfait qu'elles jaillissent toujours aussi pures et abondantes, alors que d'autres doivent recourir aux sentiers boueux d'une actualité sordide. Que si on lui fait reproche de la superbe de son attitude, si on le somme de s'engager, si on demande pourquoi il s'obstine en ces hauts glaciers ou nul que ses congénères ne peut le suivre, il répond avec Jacobi: "Pour l'honneur de l'esprit humain!"

own reasons to do mathematics. In an interview published in "Pour la Science" (November 1979, the French version of "Scientific American") he says:

> According to Plutarch, it is a noble ideal to work to make one's name immortal. Ever since I was young, I hoped that my work would have a certain place in the history of mathematics. Is that not a motivation as noble as to try to get a Nobel prize?[4]

So, it's not so much for the honor of the human spirit, it's for the honor of his own spirit. I think rather that one does mathematics because one likes to do this sort of thing, and also, much more naturally, because when you have a talent for something, usually you don't have any talent for something else, and you do whatever you have talent for, if you are lucky enough to have it. I must also add that I do mathematics also because it is difficult, and it is a very beautiful challenge for the mind. I do mathematics to prove to myself that I am capable of meeting this challenge, and win it.

So one does mathematics, but that does not mean people are unhappy if the mathematics they do is sufficiently good to make it in the history books. Of course, all the mathematicians that I know are perfectly happy when they do mathematics at this level. They are happy with the possible honors they may get from it, and they are happy to leave a name in mathematics. But I would not say that they do mathematics specifically for this purpose, that they give themselves to mathematics, whether they be pure or applied.

If I ask you what music means to you, would you answer: "The manipulation of notes"? When one does pure mathematics, one does something quite different from "manipulating". To make clear the reasons behind people doing pure mathematics, from an aesthetic point of view, I have to give you an example. But to show you what mathematics is, if you are not yourself in mathematics, I have difficulties which are analogous to those which I would have if I tried to tell an ancient Japanese, or a Hindu who never had contact with Western civilization, what a Beethoven symphony or a Chopin ballade is like. If you take someone totally foreign to Western culture, and deaf besides, how can you make that person realize what a Beethoven symphony or a Chopin ballade is like? It's impossible. Even if the person is not deaf, and is able to listen, it is still almost impossible if the person has no connection with Western culture, if the person has not heard these pieces several times. Western music is too different from Japanese music, or Hindu music; it is played on different instru-

---

[4] In a conference at the International Congress of Mathematics in Helsinki, 1976, reproduced in his *Collected Works* Vol. III, Weil had already touched this theme: "That mankind should be spurred on by the prospect of eternal fame to ever higher achievements is of course a classical theme, inherited from antiquity; we seem to have become less sensitive to it than our forefathers were, although it has perhaps not quite spent its force."

ments, with different orchestrations, with different rhythms, etc. So there is a great difficulty in making somebody understand what it's about. And conversely, Koto or Sitar concerts here in Paris don't happen so often, and affect only a small number of people.

Besides, there is a difficulty which occurs in all aesthetic situations: somebody may like one thing and not another. There are people who like Brahms and don't like Bach; who like Bach and don't like Chopin; who like Chopin and don't like Dowland (an English composer of lute pieces and lute songs at the time of Shakespeare).

How are you going to make somebody understand what a song by Dowland is like, or a Chopin ballade, without making them listen? It's impossible! And it's much easier to make you listen to some music than to make you do mathematics, because to listen to music you are in a passive state. You are taken in by the musical aesthetic, and you let the composer and interpreter take the active part. But to do mathematics, you need a much higher degree of concentration, and a personal effort. Furthermore, to make you do mathematics, I have to find a topic which is sufficiently deep, which is a real topic of mathematics, recognized as such by mathematicians. I can't cheat, but still I have to be able to explain things with words which everybody will understand. There are only very few such topics; and since I have to make a choice, maybe some people will like it and some others won't like it.

The topic has to be sufficiently deep to make you understand why some people will do mathematics all their life, and perhaps will neglect their wives, or husbands, or children, or God knows what. By the way, let me read you two sentences from a letter by Legendre to Jacobi[5] who had just gotten married rather late in life:

> Congratulations for having met a young wife who, after a *rather long* experience, you decided will make you happy forever. You were of a suitable age to get married. A man destined to spend a lot of time working in his office needs a companion who will deal with all the details of housework, and saves her husband from having to worry about those small day to day items which a man is not able to handle.

The sentence has a funny ring, especially in our "liberated" age.

Well, I have been talking in generalities for about ten minutes, that's enough. Now let's do mathematics. In the choice of the subject, I am very restricted, and it was almost necessary to pick a topic having to do with numbers. It concerns prime numbers.

Who has heard of prime numbers? [*Varied reactions and response in the audience.*] Almost everybody, or nobody? Raise your hand. Who has never heard of prime numbers? [*Almost everybody in the audience has heard about prime numbers and knows approximately what the word means.*] For instance, you, Madam, what are the prime numbers?

---

[5] Written 30 June 1832, loc. cit p. 460.

LADY. 1, 3, 5, 7 . . .

SERGE LANG. No! These are the odd numbers. I mean the prime numbers, that is 2, 3, 5, 7, 11, 13. What's the next one?

LADY. 17, 19 . . .

SERGE LANG. Very good, you have understood what a prime number is.

LADY. I forgot 2.

SERGE LANG. Yes, you are right. I misunderstood. But it is a general convention that 1 is not called a prime number. So to say that a number is prime means that it is at least equal to 2, and that it is divisible only by itself and by 1.

The number   4 is not prime because $4 = 2 \times 2$.
6 is not prime because $6 = 2 \times 3$.
8 is not prime because $8 = 2 \times 4$.
9 is not prime because $9 = 3 \times 3$.

And so on. As for the prime numbers, we have already listed them up to 19. After that, we find 23, 29, 31, 37 . . .

Now here is a question about prime numbers. Are there infinitely many of them or is there only a finite number of them?

LADY. Yes, infinitely many.

SERGE LANG. Very good. How do you prove it?

LADY. I don't know.

SERGE LANG. [*Pointing to a young man.*] You, do you know how to prove it?

YOUNG MAN. Mathematicians have found millions of them.

SERGE LANG. No, I don't mean finding millions of them, I mean prove that the sequence of prime numbers does not stop.

[*Brouhaha in the audience, various proofs are suggested by some people.*]

SERGE LANG. Are you a mathematician? Yes? OK, I ask the mathematicians in the audience not to say anything. I am not talking here for them. [*Laughter.*] Otherwise, it's cheating.

I say that there are infinitely many prime numbers. This means that the sequence of prime numbers does not stop. And I am going to prove it, because there is a very simple proof, which is also very old, and is attributed to Euclid. Here is how the Greeks did it.

Let's start with a remark. Take any integer, that is a whole number, for instance 38, which I can write as 2×19 where 2 and 19 are prime numbers. Then 38 is a product of these two prime numbers. If I take 144, then I can write

$$144 = 12 \times 12 = 3 \cdot 4 \cdot 3 \cdot 4 = 3 \cdot 2 \cdot 2 \cdot 3 \cdot 2 \cdot 2.$$

Again, it's a product of prime numbers, and I have written some of them several times. In any case, I can always express an integer as a product of prime numbers. Because if I am given an integer N bigger than 2, then either N is already prime or N can be expressed as a product of two smaller numbers. Each one of these smaller numbers is either prime, or can be expressed as a product of still smaller numbers. If you continue this process, you end up with prime numbers.

Now let's give the Greeks' proof that there are infinitely many primes. We are going to see that if we make a list of the primes

$$2, 3, 5, 7, 11, 13, 17, \ldots, P$$

going from 2 to P, then we can always find another prime number which is not in this list. We proceed as follows. I take the product of all the primes in the list. This gives me some number, to which I add 1. Let N be this new number. Thus we have

$$N = (2 \cdot 3 \cdot 5 \cdot 7 \cdot \ldots \cdot P) + 1.$$

Then either N is prime, or N is not prime. If N is prime, it is not equal to any of the ones we listed from 2 to P, and so we have just constructed a new prime number. If N is not prime, then we can express N as a product of primes. In particular, we can write $N = qN'$, where q is a prime number dividing N. Can q be equal to any one of the primes from 2 to P?

PEOPLE IN THE AUDIENCE. It's a new one.

SERGE LANG. Why? Let's pick on somebody. You, the young man over there.

YOUNG MAN. For the others, the division does not come out exact.

SERGE LANG. That's right, if we divide N by q then there is no remainder; but if we divide N by one of the primes between 2 and P there is a remainder of 1. So we discovered a new prime which was not in the list. This means that you can't make up a finite list of all the prime numbers, and this concludes the proof.

Now how are the primes distributed among all numbers? Is there some rule which tells you how many there are? How they are distributed among all integers?

A GENTLEMAN. There are millions of them.

SERGE LANG. Sure, there are also billions, but that's not the question I am raising. For example, how many primes are there smaller than 10,000, approximately? Can you answer that?

SOMEONE. You can count them.

SERGE LANG. That's true, but if I said up to 1,000,000, or up to an arbitrary number x? Let's put the question differently. Is there a formula which gives the number of primes less than x? Who says yes? An approximate formula. [*Hesitations in the audience, people comment simultaneously.*] OK, it's complicated. I would have to describe the primes more precisely. Let's not go into this right away. Let me go on to raise other types of questions about primes. In particular, what are called the twin primes.

For example:

3 and 5 differ by 2;
5 and 7 differ by 2;
11 and 13 differ by 2;
17 and 19 differ by 2;
29 and 31 also.

One says "twin primes" for obvious reasons.

Now, is there an infinite number of primes like that, an infinite number of twin primes?

Who says yes? Raise your hand. [*Some hands go up.*]

Who says no? [*Other hands go up.*]

Who keeps a prudent silence? [*Many hands go up. Smiles.*]

Who thinks it's an interesting question?

THE AUDIENCE. Yes, it's interesting. [*Several people talk at once.*]

SERGE LANG. Of course, you can like it or not like it. In fact, mathematicians generally think it's an interesting problem. Well, you see, it's a problem. No one knows the answer. If you find the answer, you will be like in Plutarch, you'll make it in the history of mathematics. In fact, one thinks there are an infinite number, and one can even do better than that. One can try to understand why there should exist an infinite number of twin primes.

SOMEONE. Is there an infinite number of triplets?

SERGE LANG. The question is interesting. Can you answer it right away?

SEVERAL VOICES IN THE AUDIENCE. Yes, I think there is an infinite number.

SERGE LANG. Watch out! Let's try to add a number to the couples of primes we already have.

3   5   7
5   7   9
11 13 15
17 19 21
29 31 33
etc.

SOMEONE. 21 is not prime.

SERGE LANG. Yes. What do you notice with your triplets? There is one, 3, 5, 7. But after that, what happens? You don't know? Look carefully: 9, 15, 21, 33 . . .

AUDIENCE. They are multiples of . . .

SERGE LANG. Shhh! The gentleman over there. [*Hesitations. No answer from the gentleman.*] They have a property, those numbers: they are all divisible by 3. That's a very easy exercise, to show that in every triplet of odd numbers, there is always a multiple of 3. Hence there cannot be a triplet of prime numbers.

AUDIENCE. Except the first one, 3, 5, 7.

SERGE LANG. Except the first, of course, which also has a multiple of 3, but 3 is prime, and there won't be any other.

Let's go back to the twin primes, the couples of primes if you want. Let's try to understand why there should be an infinite number of them. But before, let's go back to the question: how many primes are there less than or equal to x? An approximate formula.

OK, let's take all integers up to x:

$$1, 2, 3, 4, 5, 6, 7, 8, 9, \ldots, x.$$

Among these numbers, you have the even numbers and the odd numbers. What does it mean that a number is prime? It means that it is divisible only by itself and 1. Therefore, if a number if prime, it is certainly not even.

AUDIENCE. Except 2.

SERGE LANG. Of course, except 2. Now, if I go up to x, how many odd numbers are there?

SEVERAL VOICES IN THE AUDIENCE. Half of them.

SERGE LANG. Approximately half. That's right, $x/2$. It's a certain fraction of x. The number of primes less than or equal to x will be a certain fraction times x. And this fraction will depend on x. It is this fraction which we are trying to determine.

All right, so among all the integers 1, 2, 3 up to x, there will be approximately half of them which will be odd, so not divisible by 2. Among the odd numbers, how many will not be divisible by 3?

AUDIENCE. One third.

SERGE LANG. No, one third is divisible by 3 and two thirds won't be divisible by 3. OK? Let's write 2/3 in the form $(1 - 1/3)$. Now among the remaining ones, how many will not be divisible by 5?

A VOICE IN THE AUDIENCE. $1 - 1/5$.

SERGE LANG. Are you a mathematician? Yes? Then shut up! It's cheating. It's not nice. Among the remaining ones, how many are there which are not divisible by the next prime number?

AUDIENCE. $1 - 1/7$.

SERGE LANG. Good, and then finally to find the numbers which are prime, what do we need? We need that they should not be divisible by any prime number, from 2 to . . . somewhere. Thus we are led to take the product

$$\frac{1}{2}(1 - \frac{1}{3})(1 - \frac{1}{5})(1 - \frac{1}{7}) \cdots$$

which must go up to where?

AUDIENCE. Up to the last prime number before x.

SERGE LANG. Yes, but one can do better than that. Anyhow, at worst, it will be the product

$$\text{product of all factors } (1 - \frac{1}{p})$$

where p goes up to x. This will be approximately the fraction of x which gives the fraction of all numbers which are prime.

Now, in fact, I don't need to go up to x. I need to go only up to the square root of x, which is denoted by $\sqrt{x}$. Because suppose that a number which is smaller than x and is not prime, is divisible by some prime bigger than $\sqrt{x}$. Then it is necessarily divisible by a prime number smaller than $\sqrt{x}$.[6] Hence we can eliminate such a number when we have met the smallest of its prime factors. But when x is large, and when p is between $\sqrt{x}$ and x, the term $(1 - 1/p)$ is very close to 1. One can show that the product taken over all p with $\sqrt{x} \leq p \leq x$ is close to $1/2$. To simplify the formulas, I shall continue to write the product with $(1 - 1/p)$ for all $p \leq x$. To have a better approximation, or the best possible approximation, I would anyhow have to multiply the product by a constant which is hard to determine, and which reflects relations which are more hidden than the relation which we have just described.

Here I count approximately, and I am led to consider that product. It gives approximately the fraction of x giving the number of primes less than or equal to x. This fraction of x is rather mysterious, but still, it gives some idea of what's happening. For instance, is this fraction constant? Clearly not. The further we go, the smaller it becomes. If I take x very large, the fraction will be small. How fast it becomes small is not clear. It's not at all clear how this product behaves. And now, I am stuck. I will give you some answer later, but I won't be able to prove it because it would get too technical.

---

[6] I give the details of this assertion. Let N be less than or equal to x. Suppose that N is a product, $N = pN'$ with p prime greater than $\sqrt{x}$. Then $N' = N/p$, and N' is smaller than $\sqrt{x}$. If q is a prime factor of N', then q is smaller than $\sqrt{x}$ and is also a factor of N.

It's complicated to analyze this product, but still, we have made a step forward by finding this product, which gives us some fraction of x, decreasing as x increases.

Mathematicians use the sign

$$\prod$$

to denote a product. So we denote the product of all the factors $(1 - 1/p)$, taken for all primes p less than or equal to x, by the symbol

$$\prod_{p \leq x} (1 - \frac{1}{p}).$$

The number of primes $\leq x$ should then be approximately equal to

$$\prod_{p \leq x} (1 - \frac{1}{p}) \cdot x.$$

Since it's a little heavy to write the product, we are going to express it by a single letter, $F(x)$ (F for "fraction", depending on x). So we let

$$F(x) = \prod_{p \leq x} (1 - \frac{1}{p}).$$

With this abbreviation, we can then write that the number of primes $\leq x$ is approximately equal to

$$F(x)x,$$

which looks simpler.

Now, let's try to apply the same analysis to the twin primes. What happens to twin primes which did not happen for all prime numbers? There is one extra restriction: if p is prime, then $p + 2$ must also be prime. Let's take all numbers

$$1, 2, 3, 4, 5, 6, 7, 8, 9, \text{ up to } x.$$

About half of these are odd. So again we get a factor of $1/2$. Now let's look at those which are not divisible by 3, and let us write under each number the remainder after we divide it by 3:

$$
\begin{array}{ccccccccc}
1 & 2 & 3 & & 4 & 5 & 6 & & 7 & 8 & 9 \ldots \\
1 & 2 & 0 & & 1 & 2 & 0 & & 1 & 2 & 0 \ldots
\end{array}
$$

Since p cannot be divisible by 3, after we divide it by 3 we get a remainder of 1 or 2. We have two possible choices.

For the twin primes, both p and $p + 2$ must be prime. So not only p is not divisible by 3 but also $p + 2$ is not divisible by 3. This means that when we divide p by 3, the remainder must be . . .

THE AUDIENCE. Different from 1.

SERGE LANG. Yes, because if the remainder is equal to 1, and if I add 2, then $p + 2$ is divisible by 3. So we have found a new condition on p, that after dividing by 3, the remainder must be 2. Instead of excluding just one possibility, as we did before, we now exclude two possibilities. Our product therefore starts with

$$\frac{1}{2}(1 - \frac{2}{3}).$$

Now let's do the same thing with 5. If we divide an integer by 5, and the integer is not divisible by 5 exactly, then there are four possible remainders, namely 1, 2, 3, 4. Among these, if I add 2, I want that the number $p + 2$ is also not divisible by 5. Then how many possible remainders are there? In other words, in order that $p + 2$ is not divisible by 5, the remainder should not be equal to what?

AUDIENCE. 3.

SERGE LANG. Yes indeed, if we divide the integer by 5, the remainder should be different from 0 or 3. This gives me a factor

$$\frac{3}{5} \quad \text{or} \quad (1 - \frac{2}{5}).$$

Next for 7, I want to characterize those integers p which are not divisible by 7, and such that, if I add 2, then $p + 2$ is not divisible by 7. Then I must omit multiples of 7, and in addition those whose remainder after dividing by 7 is equal to 5. The next factor will therefore be . . .

AUDIENCE. $(1 - 2/7)$.

SERGE LANG. Excellent. Therefore the fraction we are looking for will be the product

$$\frac{1}{2} \prod (1 - \frac{2}{p}),$$

taken over all prime numbers ≥ 3 and less than or equal to x. When we considered all prime numbers, without any further restriction, we were led to take the product of all terms $(1 - 1/p)$. Now with the extra condition that $p + 2$ is prime, we are led to the product of the terms $(1 - 2/p)$. All of this is approximate, but it gives a good idea about how many twin primes there are. That's the conjecture:

Conjecture. The number of twin primes less than or equal to x is approximately equal to

$$\frac{1}{2} \prod_{3 \leq p \leq x} (1 - \frac{2}{p}) x.$$

Here again, the product changes with x, it is a function of x. It is not a constant function like $4/5$, or $1/12$. As before, we abbreviate the product, and we let

$$F_2(x) = \frac{1}{2} \prod_{3 \le p \le x} (1 - \frac{2}{p}),$$

so that the number of twin primes $\le x$ is approximately equal to $F_2(x)x$. We are now in a similar situation as when we were counting all the prime numbers, and there remains to analyze this product, which is taken over prime numbers even though we are trying to count prime numbers. There is something a little circular here, but not completely.

We get some information from this product. One can compute this product. Even though the fraction

$$\frac{1}{2} \prod_{3 \le p \le x} (1 - \frac{2}{p})$$

decreases with x, this fraction is still rather large, but I would have to explain what I mean by "rather large". Now I am stuck, one can say it only with some more advanced vocabulary, with slightly more knowledge of mathematics. Up to now, I could manage only with the basic rules of arithmetic that one uses in the 7th grade. But let's try anyhow.

Who has heard of the logarithm? [*A few hands go up.*] Who never heard of the logarithm? [*A few hands go up.*] Who keeps a prudent silence? [*Several hands go up.*] OK, there is something that's called the logarithm. It is denoted by $\log x$. You will find it on all the little hand calculators in the drug stores. I don't have time to explain it in greater detail. [*A few more explanations are given later.*]

Then it is true that

$$\prod_{p \le x} (1 - \frac{1}{p}) \text{ is approximately equal to } \frac{1}{\log x}.$$

But this is not trivial to prove, and there is no way I can give you any idea how it is proved. It's quite technical, and it's even tough to do. It's elementary if you start from differential and integral calculus, but even being elementary, it's tough. You might manage in say . . . thirty pages.

[*Various reactions in the audience.*]

SERGE LANG. Oh, you know, thirty pages, it's nothing. Six months ago some new theorems got proved that required 10,000 pages. So thirty pages, it's no big deal. Starting from scratch, of course.

Anyhow, there is a function which is called $\log x$, and the first product

$$\prod_{p \le x} (1 - \frac{1}{p})$$

is approximately equal to $1/\log x$.

As for the other product, associated with the twin primes, one can prove that

$$F_2(x) \text{ is approximately equal to } \frac{1}{(\log x)^2}.$$

The square comes from the fact that we replace $1/p$ by $2/p$. For example, we have

$$(1 - \frac{1}{p})^2 = 1 - \frac{2}{p} + \frac{1}{p^2},$$

and if p is large, then $1/p^2$ is very small compared to $2/p$. So approximately we can leave it out, and we find that

$$\prod (1 - \frac{1}{p})^2 \text{ is approximately equal to } \prod (1 - \frac{2}{p}).$$

Therefore, the conjecture is:

The number of twin primes less than or equal to x is approximately equal to

$$F_2(x)x, \quad \text{or also to } \frac{x}{(\log x)^2}.$$

Naturally, I would still have to explain more precisely what I mean by "approximately", and I don't have the time now to do so. It is a little more technical. Maybe we will have time later, after the talk.

The function $\log x$ is a function which grows slowly with x. Therefore our fraction is relatively large. But in spite of these heuristic arguments, nobody knows how to prove that there exists an infinite number of twin primes.

What have I just done? There is no doubt that we have been doing mathematics! But nothing has been proved, except the first theorem of Euclid. We have given arguments which were only heuristic, but that does not mean that the mind did not function. On the contrary. We formulated a conjecture, which means that we tried to guess what was the answer, and we now face a problem. Well, that's what it means to do mathematics: find interesting problems and try to solve them. Eventually, solve them.

Now let's raise another question. We observe that:

$2^2 + 1 = \quad 4 + 1 = \quad 5$ is prime
$4^2 + 1 = \quad 16 + 1 = \quad 17$ is prime
$6^2 + 1 = \quad 36 + 1 = \quad 37$ is prime
$8^2 + 1 = \quad 64 + 1 = \quad 65$ is not prime
$10^2 + 1 = 101$ looks like it should be prime; in fact it is prime.

Question: In this list of prime numbers which can be written as the square of a number plus one, are there infinitely many primes? Think about it, I am asking for your intuition. I am not asking you to prove anything yet. Are there infinitely many primes of the form $n^2 + 1$?

SOMEBODY. No.

SERGE LANG. Who says yes . . . ? Who says no . . . ? Who keeps a prudent silence? [*Varied reactions in the audience. Guesses go both ways.*] It's less clear, isn't it?

AUDIENCE. There is more space between them. They occur less frequently.

SERGE LANG. That's right, madam, there is more space between them. And there is more space than there was between the twin primes, which in turn had more space between them than all the primes. Can we guess how much space there should be, approximately? A little? A lot? Can you give a quantitative measure?

First let me give you the answer: nobody knows if there exists an infinite number. It's an unsolved problem. It's one of the great problems of mathematics. One thinks that the answer is yes. I repeat, if you find the answer, you will make it into the history books of mathematics (but you didn't necessarily do it with that purpose in mind).

The conjecture is that there exists an infinite number of primes of the form $n^2 + 1$, but like for the twin primes, one can do better than that. We can give some idea of the corresponding fraction that they represent.

For all the primes, the fraction is

$$F(x)x \quad \text{or} \quad \frac{1}{\log x} \, x = \frac{x}{\log x}.$$

For the twin primes, the fraction is

$$F_2(x)x \quad \text{or} \quad \frac{1}{(\log x)^2} \, x = \frac{x}{(\log x)^2}.$$

What fraction are we going to find for the primes of the form $n^2 + 1$?

SOMEONE. You must necessarily have n smaller than $\sqrt{x}$.

SERGE LANG. Right! If $n^2 + 1$ is smaller than x then n is bounded by $\sqrt{x}$. Let's try to guess what fraction of all numbers is represented by the primes of the form $n^2 + 1$. If the primes are distributed at random, it is probably the same fraction of $\sqrt{x}$ as the fraction of all primes with respect to x. It's rather plausible. Anyway, it's a working hypothesis. So what is the conjecture? The gentleman over there.

GENTLEMAN AND THE AUDIENCE. [*Everyone hesitates.*]

SERGE LANG. The fraction of those primes less than or equal to x is

$$\frac{1}{\log x} \, x.$$

If you apply this to $\sqrt{x}$ you get approximately

$$\frac{1}{\log x} \; \sqrt{x}\,.$$

That's the conjecture, roughly speaking, up to a constant factor.

SOMEONE. Why not $\dfrac{1}{\log \sqrt{x}} \; \sqrt{x}$?

SERGE LANG. OK, it's not so clear if it should be x or $\sqrt{x}$. But first, one has the relation

$$\log \sqrt{x} \;=\; \frac{1}{2} \log x,$$

so the two expressions differ only by a factor of 2; and second, I don't claim to give anything but an approximation, up to some constant factor. In any case, these heuristic ideas, which are purely intuitive, give you the idea that there should be an infinite number of such primes, since one can give a quantitative measure for them.

Of course, I should explain what I mean by "approximately", not only for the primes of the form $n^2 + 1$, but also for all the primes, or the twin primes. This would be the topic for another talk, which I can't give today and which would last perhaps one hour. It's precisely the error term in this approximation that is the subject of a problem which is generally recognized as being the greatest problem in mathematics. It's the error term which appears in the formula $x/(\log x)$ for all the prime numbers. There is a precise conjecture, due to Riemann, and called the Riemann hypothesis, made about 130 years ago, and which gives the best possible error term. It's still not proved today, despite the fact that many mathematicians have worked on it.

But I have been talking for an hour. Let's stop here.

The questions

QUESTION. You have mentioned other pure mathematicians, but you, why do you do this kind of work?

SERGE LANG. Why? Why do you compose a symphony or a ballad? I already told you why. Because it gives me chills in the spine. That's why. But I did not say you should also get them. That's freedom for you.

QUESTION. Can you say where is the limit between pure and applied mathematics?

SERGE LANG. There are no limits. The two mix with each other without my being able to define a limit. If you try to define a limit more precisely, in general, I don't say you won't succeed, but I have personally never seen anyone succeed in doing so.

QUESTION. What you did just now, do you think it could be useful somewhere?

SERGE LANG. You said "could". This is a conditional, so I am forced to answer logically: yes.

QUESTION. When you do mathematical research, do you have a goal in mind?

SERGE LANG. The goal is to prove the conjecture.

QUESTION. But at the start?

SERGE LANG. At the start, it's first to find the conjecture that you want to prove, and then try to prove it. One of the main difficulties in mathematics is to find the subject on which you want to concentrate, and the problem which you are going to try to solve.

QUESTION. But is that done by logical deduction or intuition?

SERGE LANG. Have I done any logic here? Half and half. There was a lot of intuitive stuff, and logic, you know, when I tell you that something or other is one third or one fifth of something else, I have assumed a lot of things without proving them. It's more by intuition than by logic that I have been doing mathematics here. Anyway, in general, new results are discovered by intuition, proofs are discovered by intuition, and finally they are written up according to a logical pattern. But don't confuse the two. It's the same as in literature: grammar and syntax are not literature. When you write a musical piece, you use notes, but the notes are not the music. To read a piece of music from the written text is not a substitute for hearing the piece in Carnegie Hall or elsewhere. Logic is the hygiene of mathematics, just as grammar and syntax are the hygiene of language—and even then! "Under the bam, under the boo, under the bamboo tree . . .", there isn't any grammar. The essential thing in Shakespeare, or Goethe, is not grammar or syntax. It is the poetry, the musical effect of words, poetic allusions, aesthetic impressionism, and many other things. But whereas the beauty of poetry pales under translation, the beauty of mathematics is invariant under linguistic transformations.

QUESTION. You have used heuristic arguments, and approximations to describe what a pure mathematician does. But a mathematician does other things besides that.

SERGE LANG. Watch out, I did not say that a mathematician does only that. One tries to prove something, one discovers a conjecture a little like I have described here. But once the conjecture has been made, one tries to prove it. Sometimes we succeed, sometimes we don't. We proceed by successive approximations, both in making guesses and trying to prove them. The negation of one absolute is not the absolute of opposite type.

Depending on how often you succeed, or how deep are your results, you will be a great mathematician, or an average one, or . . .

QUESTION. For instance, you haven't talked about axiomatization.

SERGE LANG. Axiomatization is what one does last, it's rubbish. It's the hygiene of mathematics, axiomatization. It's the discipline of the mind. Like grammar and syntax. But do what you want. Each one has to determine what they like to do. The word "rubbish" is too strong. I also axiomatize, when I find it appropriate to do so, and there are lots of other things I have not talked about. I made a choice. I wanted to show an essential aspect of mathematics which most people have no idea exists.

SOMEONE. There is a problem that gives me chills in the spine, the problem of the denumerability of the real numbers. Cantor tried to deal with this problem, and I think he became a little crazy because of it. I have heard that Cantor proved it. I'd like to know if this is true.

SERGE LANG. Proved what? that the real numbers are not denumerable? Yes, he surely did.

SAME. Can you give us an idea of the proof?

SERGE LANG. [*Hesitates.*]

SAME. Without going too far.

SERGE LANG. OK, the gentleman would like . . . [*Brouhaha in the audience.*]

Yes! I can do it in just a few minutes.

GENTLEMAN. I was just curious.

SERGE LANG. But that's all it ever is, curiosity! [*Laughter.*] On the contrary, the whole point of the operation was to sharpen your curiosity by showing you what I was curious about. So I give the proof. What is a real number? It's an infinite decimal, for example 27.9130523 . . . Since I can't write an infinite number of digits like that, I have to use some notation with indices. And to simplify matters, I'll consider only the numbers between 0 and 1. Suppose that we can write all these numbers in a sequence, with a first, a second, a third, and so on, without missing any of them, as follows:

$$0.a_{11}\, a_{12}\, a_{13}\, a_{14} \ldots$$

$$0.a_{21}\, a_{22}\, a_{23}\, a_{24} \ldots$$

$$0.a_{31}\, a_{32}\, a_{33}\, a_{34} \ldots$$

with integers a_{ij} between 0 and 9. I am going to show that there is some infinite decimal which is not in this list. I choose an integer b_1 which is not equal to a_{11}. Then an integer b_2 which is not equal to a_{22}. Then an integer b_3 which is not equal to a_{33}. In general, I choose an integer b_n which is not equal to a_{nn}, and I pick b_n between 1 and 8 (to avoid ambiguities having to do with a sequence of 0's or 9's). Then the infinite decimal

$$0.b_1\, b_2\, b_3\, b_4 \ldots$$

is not equal to any decimal in the list because of the way I have constructed it, so it is a new one.

Note that what we have just done is similar to Euclid's method at the beginning. We made a list, and then we showed that there is a decimal which is not in the list.

QUESTION. I would like to know what you think of the great schools of mathematical thought concerning infinity.

SERGE LANG. I don't think about it. All of this was settled for me long ago. It had some historical importance, but today, it's settled. Something is either infinite or it is not.

QUESTION. But it's not as simple as that!

SERGE LANG. OK, you are right.

QUESTION. Does infinity exist?

SERGE LANG. When I mentioned prime numbers, did you know how to answer whether there was an infinite number of them or not?

QUESTION. Yes.

SERGE LANG. Then that's it, you have understood. That settles the question.

QUESTION. But Cantor's proof was more or less rejected by the intuitionists. I think there was a lot of fighting about this subject.

SERGE LANG. If people want to fight, they are free to do so. I just do mathematics.

QUESTION. Have you worked yourself on the problems you raised today?

SERGE LANG. Yes, on the problem of primes of the form $n^2 + 1$. Since that interests you, and you are still sitting here, let me give a few more precise statements about that problem. When I started to think about what I would tell you today, I thought of the twin primes but I did not know myself if there was a conjecture about them, nor how to motivate it. I looked up the book by Hardy and Wright, and I found it. This conjecture, and the one about the primes of the form $n^2 + 1$ are due to Hardy and Littlewood, in an article dating back to 1923. I am going to state their conjecture somewhat more precisely than I have done so far.

I have said several times that certain expressions were approximate, up to a constant factor. What does this mean? Suppose I have two expressions $A(x)$ and $B(x)$. We say that $A(x)$ is asymptotic to $B(x)$ if the quotient

$$\frac{A(x)}{B(x)}$$

approaches 1 when x grows larger and larger. This means that when x is very large, then the quotient is very close to 1. The relations that $A(x)$ is asymptotic to $B(x)$ is denoted by the symbol

$$A(x) \sim B(x).$$

We can then state the prime number theorem as follows.

Let $\pi(x)$ be the number of primes $\leq x$. Then we have the relation

$$\pi(x) \sim e^{\gamma} F(x)x,$$

where e and γ are constants used all the time in mathematics and F is as before. The constant e is called the natural base for logarithms; and γ is called Euler's constant. Since the product $F(x)$ itself looks rather mysterious, one prefers to replace it by another expression. It is a theorem due to Mertens that one has the asymptotic relation

$$e^{\gamma} F(x) \sim \frac{1}{\log x},$$

and therefore we find that

$$\pi(x) \sim \frac{x}{\log x},$$

which is the usual formulation for the prime number theorem. It is useful to write it this way, because the log function is very well known. We know how it grows when x becomes large. For instance, we have the following values:

$\log 10 = 2.3\ldots$	$\log 10,000 = 9.2\ldots$
$\log 100 = 4.6\ldots$	$\log 100,000 = 11.5\ldots$
$\log 1000 = 6.9\ldots$	$\log 1,000,000 = 13.8\ldots$

and so on. Observe that the numbers 10, 100, 1,000, 10,000, 100,000, 1,000,000 grow by powers of 10, but the logarithm grows only by adding approximately 2.3 each time. This means that the logarithm grows much more slowly.

Similarly, let $\pi_2(x)$ denote the number of twin primes $\leq x$. Then Hardy–Littlewood's conjecture is that

$$\pi_2(x) \sim (e^{\gamma})^2 F_2(x)x.$$

This formula can also be written asymptotically with the logarithm, in the form

$$\pi_2(x) \sim 2C_2 \frac{x}{(\log x)^2},$$

where C_2 is a constant, given by an infinite product taken over all primes ≥ 3, namely

$$C_2 = \prod_{3 \leq p} \left[1 - \frac{1}{(p-1)^2} \right].$$

Hardy and Littlewood give probabilistic arguments more precise than those I could give here in one hour. In particular, when I wrote the products, I was assuming implicitly that the conditions of divisibility by 2, 3, 5, etc. were independent. But I did not prove this assumption, which in fact is false. These conditions are not independent, and the constant e^γ reflects the dependencies between these divisibility conditions.[7] But this is now getting much more technical, and I cannot go into the details necessary to find the constant e^γ. I have to refer you to the original article by Hardy–Littlewood, or the book by Hardy and Wright.

To come back to the question about my own work, I and a friend Hale Trotter have been interested in analogous problems, concerning the distribution of prime numbers in much more complicated contexts. I can't go into them here. But we rediscovered the same asymptotic relation as Hardy–Littlewood for the primes of the form $n^2 + 1$, with the same constant C_2 (fortunately!). The article with Trotter gives a probabilistic model which is completely different from that of Hardy–Littlewood. Naturally, only someone who has specialized in number theory can understand it.

QUESTION. Between pure and applied mathematics, I don't see the difference very well.

SERGE LANG. At first sight, to compute the number of primes of the form $n^2 + 1$ has no applications. This does not mean that it will never have applications. In the history of mathematics, the results of research done purely from an aesthetic point of view have been applied, sometimes after a century, to very concrete problems. For instance, today, one uses parts of the theory of prime numbers in coding theory. As far as I know, it's not the same theorems that we have discussed today, but it could very well be.

I have also brought a quote from von Neumann,[8] which I did not have the time to read before. Maybe it's time to read it now. [*Approval from the audience.*] OK, here it is.

---

[7] The proof of the conjectured formula for the number of primes is not at all trivial. Indeed, the Goldbach problem, which is entirely analogous to the twin prime problem, states that every sufficiently large even number is the sum of two odd primes. Hardy and Littlewood have even conjectured that there is an asymptotic formula for the number of such representations, given by

$$N_2(n) \sim 2C_2 \frac{n}{(\log n)^2} \prod \frac{p-1}{p-2},$$

where the product (finite) is taken over all the primes $\neq 2$ dividing n. Note again the same constant C_2 which we found in the twin prime problem, as well as the denominator with the square of the logarithm. The heuristic arguments are similar. But Hardy–Littlewood remark that Sylvester in 1871 and Brun in 1915 had conjectured a false formula, which did not take into account the relations giving rise to the factor e^γ.

[8] J. von Neumann, *The Mathematician*, Collected Works I, pp. 1–9.

I think it is a relatively good approximation to truth—which is much too complicated to allow anything but approximations—that mathematical ideas originate in empirics, although the genealogy is sometimes long and obscure. But, once they are so conceived, the subject begins to live a peculiar life of its own and is better compared to a creative one, governed almost entirely by aesthetical motivations, than to anything else, and in particular, to an empirical science. There is, however, a further point which, I believe, needs stressing. As a mathematical discipline travels far from its empirical source, or still more, if it is a second and third generation only indirectly inspired by ideas coming from "reality," it is beset with very grave dangers. It becomes more and more purely aestheticizing, more and more purely *l'art pour l'art*. This need not be bad, if the field is surrounded by correlated subjects, which still have closer empirical connections, or if the discipline is under the influence of men with an exceptionally well developed taste. But there is a grave danger that the subject will develop along the line of least resistance, that the stream, so far from its source, will separate into a multitude of insignificant branches, and that the discipline will become a disorganized mass of details and complexities. In other words, at a great distance from its empirical source, or after much "abstract" inbreeding, a mathematical subject is in danger of degeneration. At the inception the style is usually classical; when it shows signs of becoming baroque, then the danger signal is up. It would be easy to give examples, to trace specific evolutions into the baroque and the very high baroque, but this, again, would be too technical.

I have some objections to the way von Neumann expresses himself. If he expresses merely his personal tastes, well and good. He has the right to his own tastes. Unlike him, I don't feel any danger about doing mathematics for which I see no relation with the empirical world. Many times during the course of my life, I have seen situations when some mathematicians complained that certain fields of research were too "abstract"—von Neumann might say "baroque". But fifteen years later, such research combined with other led to the solution of very classical problems, which had been raised already in the 19th century.

There are as many possibilities to do uninteresting or trivial mathematics in number theory as there are doing mathematics with empirical connections. As for "inbreeding", I don't understand what von Neumann means. Many of the most beautiful discoveries in mathematics come from the wedding of branches which a priori seem very far apart from each other. One of the characteristics of mathematical genius is the ability to bring together different branches, by what could be called "inbreeding", or to bring together threads going off into many directions; to find fundamental ideas in the mass of details and complexities which others have accumulated. This does not mean that the work of others has been worthless.

Historically, in the 50's, it is true that several branches of pure mathematics developed parallel to each other. Von Neumann was not the only one to complain that these streams, which for many at the time seemed without connection to each other, were too abstract. But in the 60's, we have seen these streams come together in some very deep and essential ways. And not only that, but we have seen them join with subjects which had not been fashionable for forty years, and we have seen them join with subjects which had been almost forgotten since the 19th century. We have also seen old conjectures proved precisely because in the last fifteen years, people have found how to make syntheses which rank among the most successful in the history of mathematics. A posteriori, we see today that the parallel developments of the fifties were an essential step for the syntheses which followed.

GENTLEMAN. To return to prime numbers, we accept that there is an infinite number of them, and consequently there is an infinite number of inverses of these primes. Is it true that the sum of these inverses is finite?

SERGE LANG. That's a very nice question! You want to take the sum

$$\frac{1}{2} + \frac{1}{3} + \frac{1}{5} + \frac{1}{7} + \frac{1}{11} + \cdots$$

GENTLEMAN. Yes.

SERGE LANG. So it's the sum

$$\sum_{p \leq x} \frac{1}{p}.$$

Well, if I wanted to plant someone in the audience to ask a question which fit exactly with what I said before, I could not have done better than to have planted the gentleman over there. [*Laughter.*]

Remember that our product was

$$\prod_{p \leq x} (1 - \frac{1}{p}).$$

We have just written a sum with $1/p$. The two look like each other. One of them looks multiplicative, and the other looks additive. but the fact that they look alike is not at all accidental, and is due precisely to the logarithm which I did not have time to discuss much. But if you give me two minutes . . . The logarithm has two simple properties. The first is that

$$\log (ab) = \log a + \log b.$$

In other words, the logarithm of a product is equal to the sum of the logs. If you know the logarithm, you know this property.

The second property is that when t is very small, then $\log (1+t)$ is approximately equal to t. Therefore $\log (1-t)$ is approximately equal to $-t$.

Now suppose that I take the logarithm of the product. Since the log of a product is equal to the sum of the logs, we have

$$\log \prod_{p \leq x} (1 - \frac{1}{p}) = \sum_{p \leq x} \log (1 - \frac{1}{p}).$$

But $\log(1 - 1/p)$ is approximately equal to $-1/p$. Hence our sum is approximately equal to

$$\sum_{p \leq x} \log (1 - \frac{1}{p}) \sim - \sum_{p \leq x} \frac{1}{p},$$

which is precisely the sum which the gentleman wants to consider. It is a theorem which ones proves when you analyze the sum that we have the asymptotic relation

$$\sum_{p \leq x} \frac{1}{p} \sim \log \log x.$$

Since the logarithm grows very slowly, the iterated logarithm $\log \log x$ grows even more slowly. But it grows, and the sum is very interesting. So it is not true that the sum of the inverses $1/p$, taken for all primes p, is finite.

You see, if you study that sum, you find $\log \log x$. To study the product, you perform the inverse operation, you exponentiate, and you find $\log x$, always with a minus sign. So you find that

$$\prod_{p \leq x} (1 - \frac{1}{p}) \text{ is approximately equal to } \frac{1}{\log x},$$

which is precisely what we had before. All of this belongs to the same circle of ideas. The gentleman gets an A+.

QUESTION. Do you see applications of prime number theory in the sciences?

SERGE LANG. The sciences? You mean physics, chemistry, biology? I don't know any, but the history of mathematics shows that subjects which were considered pure can, at any moment, have the most unexpected concrete applications. I cannot predict in advance what will happen. I don't know any, but that does not mean that there aren't any, because I know practically nothing about physics and chemistry. There may be applications which I don't know about. On the other hand, I can't predict that there won't be any, and in fact, I do exactly the opposite: I say that there may be some, at any time. For instance, these last few years, pure mathematical theories in differential geometry or topology which were discovered ten or twenty years ago suddenly found applications to the theory of elementary particles in physics!

I try to avoid absolutes, from one side or another. I have told you what I like, I show you what I like. And I hope that you like it. And if it works like that, it's all I wanted to do.

Addendum

QUESTION. And the Riemann hypothesis, which you mentioned before. Can you tell us what it is?

SERGE LANG. Yes. We want to give a more precise description of the error term in the formula for the number of primes. The term $x/\log x$ is only a very gross approximation, even asymptotically. There is another expression which gives a much better approximation.

Remember that we had found a certain fraction

$$e^{\gamma} F(x), \quad \text{or also} \quad \frac{1}{\log x}$$

which we shall now call the density of primes, or also the probability that x is prime, asymptotically. After that, we said that $\pi(x)$ is asymptotic to the product of this density with x, that is

$$\pi(x) \sim \frac{1}{\log x} x.$$

But we can do better than to take this product, because $\log x$ varies with x. We get a much better formula by taking the sum of the densities, from 2 to x, which we denote by $L(x)$. That means we let

$$L(x) = \frac{1}{\log 2} + \frac{1}{\log 3} + \frac{1}{\log 4} + \frac{1}{\log 5} + \cdots + \frac{1}{\log x}$$

$$= \sum_{n=2}^{x} \frac{1}{\log n}.$$

Then we also have the asymptotic relation

$$\pi(x) \sim L(x) \sim \frac{x}{\log x},$$

but $L(x)$ gives a much better approximation of $\pi(x)$ than $x/\log x$. The Riemann Hypothesis states that

$$\pi(x) = L(x) + O(\sqrt{x} \, \log x),$$

where $O(\sqrt{x} \, \log x)$ is an error term, bounded by $C\sqrt{x} \, \log x$, where C is some constant. Since $\sqrt{x}$ and $\log x$ are very small compared to x, we see that $L(x)$ gives a very good approximation to $\pi(x)$.

The Riemann Hypothesis also allows us to understand better the relation between the product $F(x)$ and $1/\log x$. Indeed, H. Montgomery tells me that it implies the relation

$$e^{\gamma} F(x)x = \frac{x}{\log x} + O(\sqrt{x}),$$

where again $O(\sqrt{x})$ is an error term bounded by $C\sqrt{x}$, with some suitable constant C. Hence the expressions $e^{\gamma} F(x)x$ and $x/\log x$ give about the same approximation to $\pi(x)$, and both are worse than $L(x)$.

Bibliography

V. BRUN, "Über das Goldbachsche Gesetz und die Anzahl der Primzahlpaare," *Archiv for Mathematik* (Christiania) **34** Part 2 (1915), pp. 1–15.

G.H. HARDY, *A Mathematician's Apology*, Cambridge University Press, 1969.

G.H. HARDY and J.E. LITTLEWOOD, "Some problems of Partitio Numerorum," *Acta Math.* **44** (1923), pp. 1–70.

G.H. HARDY and E.M. WRIGHT, *An Introduction to the Theory of Numbers*, Fourth Edition, Oxford University Press, 1980.

A.E. INGHAM, *The Distribution of Prime Numbers*, Hafner Publishing Company, New York, 1971 (Reprinted from Cambridge University Press).

S. LANG and H. TROTTER, *Frobenius Distributions in GL_2-extensions*, Springer Lecture Notes 504, Springer-Verlag, New York, 1976.

J.J. SYLVESTER, "On the partition of an even number into two prime numbers," *Nature*, **55** (1896–1897), pp. 196–197 (= *Math. Papers* **4**, pp. 734–737).

D. ZAGIER, "The first 50 million prime numbers," *Mathematical Intelligencer*, 1978.

A lively activity: To do mathematics

Diophantine equations

15 May 1982

Summary: *Interest in solving equations in integers or rational numbers dates back from antiquity. I tried to show some fundamental problems which are still unsolved. Euclid and Diophantus already solved the equation $a^2 + b^2 = c^2$, and gave a formula for all the solutions. The next hardest equation like $y^2 = x^3 + ax + b$ has given rise to very great problems which have been at the center of mathematics since the 19th century. No one knows how to give an effective method for finding all solutions. I described some of the structures which the solutions have, and the context in which one would like to find such a method.*

In May 1981, during a brief stay in Paris, Serge Lang gave us a conference on prime numbers, showing some of the motivation which leads a mathematician to "do mathematics".

The welcome given him by the audience, the curiosity and enthusiasm of certain students who had attended his talk led him to renew the experience this year, and we are all grateful to him for doing so.

The following text was written in the same spirit as the one last year; that is, to preserve as far as possible Serge Lang's lively tone and style. The text reflects the exchange, with one deletion and a few additions. The deletion concerned an exchange on problems of high school teaching. It was either too general, or on the contrary too personal, and the questions did not seem to shed any light on this topic, so we decided to delete it. On the other hand, since Serge Lang prefers to *do* things rather than to talk about "what could be done?", the reader who would like to know more precisely how he conceives a mathematics book at this level can consult his *Basic Mathematics*,* or the book *Geometry*, written together with Gene Murrow.†

The additions deal with certain mathematical points which could not be discussed for lack of time. These points illustrate, in a certain way, the patience and kindness with which Serge Lang, in the following weeks, accepted to answer all my questions, including those which today appear rather naive. I take this occasion to express my thanks.

The last pages of the conference, concerning some conjectures about the size of solutions, were added six months later, and showed at the time, if it was still necessary, that the conference dealt with live mathematics, mathematics in the process of being done. Since then, one could not have found a better proof of the vitality and relevance of mathematical research: Mordell's conjecture (p. 55) which was about sixty years old, was proved by Gerd Faltings in Germany. This first rate result was obtained in part by using the vast resources of algebraic geometry, developed mainly during these last thirty years; and in part by relying on the work of the Soviet school of mathematics. This is a relatively frequent situation in mathematics, when a great personal contribution takes place in the context of the work developed by an active mathematical community.

J.B.

---

*Addison-Wesley, 1971 (out of print).

†Springer-Verlag, 1983.

The conference

SERGE LANG. The goal of this talk is again to do mathematics together. For those who were not here last year, I'll start with a few minutes of more general comments. Last time, I asked: "What does mathematics mean to you?" And some people answered: "The manipulations of numbers, the manipulation of structures." And if I had asked what music means to you, would you have answered: "The manipulation of notes?" So I ask again: What does mathematics mean to you?

GENTLEMAN. It's to work with numbers.

SERGE LANG. No, no! It's not to work with numbers.

A HIGH SCHOOL STUDENT. It's to solve a problem.

SERGE LANG. There, you are getting closer. Solve a problem. That's what I had tried to show you last time. That it was not just to manipulate something. It strikes much more deeply into our psychology, and unfortunately there is nothing, or almost nothing, except for certain exceptionally gifted teachers, there is nothing in our elementary schools or high schools which allows people to realize what mathematics is about, or what it means to do mathematics. Just before the conference, I was looking at a tenth grade textbook in Mr. Brette's office (he organized this conference), and it's to vomit. [*Whisperings in the audience.*] It's to vomit, from all points of view: the general incoherence, which goes from beginning to end; the little problems which don't mean anything; the aridity of the exposition . . . It's disgusting. [*Agitation in the audience, some laughter.*]

QUESTION. Can you tell us the name of the book?

SERGE LANG. Oh! I could have brought it down here, I wouldn't mind! You know, I'm not afraid to say what I think. But I left it upstairs. Anyhow, these things are practically all alike. [*Laughter.*] You know, those things are homogeneous. So what I am trying to do now, is to show you something else; to show you why mathematicians do mathematics, and spend their life doing it. That's what I am trying to show you.

Last time, we also talked about the role of pure and applied mathematics, of the relations between them, very briefly. And I read a quote from von Neumann, when he complained about what he called "baroque" mathematics. He said:

As a mathematical discipline travels far from its empirical sources, or still more, if it is a second and third generation only indirectly inspired by ideas coming from "reality", it is beset with very grave dangers. It becomes more and more purely aestheticizing, more and more *l'art pour l'art* . . . at a great distance from its empirical source . . . a mathematical subject is in danger of degeneration.

So he was complaining. But there is another quote from von Neumann which one should read to those people who pester us with the first, and don't know or don't mention the second. I am going to read it to you.

But still a large part of mathematics which became useful developed with absolutely no desire to be useful, and in situations where nobody could possibly know in what area it would become useful; and there were no general indications that it ever would be so. By and large it is uniformly true in mathematics that there is a time lapse between a mathematical discovery and the moment when it is useful; and that this lapse of time can be anything from thirty to a hundred years, in some cases even more; and that the whole system seems to function without any direction, without any reference to usefulness . . . This is true for all of science. Successes were largely due to forgetting completely about what one ultimately wanted, or whether one wanted anything ultimately; in refusing to investigate things which profit, and in relying solely on guidance by criteria of intellectual elegance; it was by following this rule that one actually got ahead in the long run, much better than any strictly utilitarian course would have permitted.

I think that this phenomenon could be studied very well in mathematics; and I think everyone in science is in a very good position to satisfy himself as to the validity of these views. And I think it extremely instructive to watch the role of science in everyday life, and to note how in this area the principle of *Laissez faire* has led to strange and wonderful results. [vN]

There is nothing like saying contradictory things to be always right. [*Laughter.*]

OK, that's enough general comments, let's do mathematics.

Of course, as I said last year, I am forced to choose topics which are in principle understandable by everybody. This means that most of mathematics is completely excluded. And it is also true that there will be numbers in the subject I have chosen for today. But it is not so much the presence of numbers that counts, as the way we are going to deal with them and think about them.

We can start without numbers, just as Pythagoras would have done, by taking a right triangle, with sides a, b, c. I suppose that everybody remembers Pythagoras' theorem, which says what? [*Serge Lang points to a young man in the audience. Laughter.*]

YOUNG MAN. The sum of the squares . . .

SERGE LANG. Yes, so what is the first square? It's a . . .

YOUNG MAN. a squared plus b squared equals c squared.

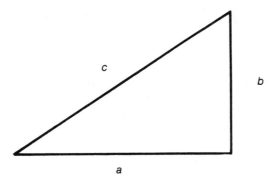

SERGE LANG. That's right, it's the equation

$$a^2 + b^2 = c^2.$$

Now, do you know any solutions of this equation in integers? Everybody knows what an integer is? 1, 2, 3, 4, 5, 6 and so on. So are there solutions with integers?

THE AUDIENCE. 3, 4, 5.

SERGE LANG. No, wait! I am asking the guy here. [*Laughter.*] Let me choose. [*Laughter again.*] And especially, the rules of the game: there are probably, and even certainly, a number of mathematicians in the audience. I ask them not to intervene, it's not for them that I am giving this talk and if they intervene, it's cheating! All right, let's go back to the young man over here. Give me a solution.

YOUNG MAN. 3 squared plus 4 squared equals 5 squared.

SERGE LANG. Yes. Now is there another one? Well, let's take a vote, we do this very democratically. You, sir, you say no. The gentleman over there thinks the answer is yes. Who says no? Raise your hand. Who says yes? There is quite a lot of yes. Those who say yes, give me another solution. Sir?

THE GENTLEMAN. [*No answer.*]

SERGE LANG. You said yes.

GENTLEMAN. I know that there are many other solutions, but it is a little difficult to say which ones.

SERGE LANG. All right, is there any one who knows another one?

THE AUDIENCE. 5, 12, 13.

SERGE LANG. It works, 25 + 144 = 169.

A HIGH SCHOOL STUDENT. If you have one, (a, b, c), and if d is any number, then (da, db, dc) will also work.

SERGE LANG. Right, if (a, b, c) is a solution and if you multiply by an integer d, then you get another solution:

$$(da)^2 + (db)^2 = (dc)^2.$$

Therefore, the reasonable question is: are there other solutions besides the two we already know, and their multiples?

Who says yes? Who says no? Who keeps a prudent silence? [*Laughter.*] In any case, we are facing a problem which the Greeks already knew. Well, what we are going to do in the next five or ten minutes, is to find all the solutions, and I will prove it. How do I prove it? I write them all down. But since I can't write them down one after another, because there is an infinite number of them, I must have a general method. So we begin by transforming the problem a little. If I divide the equation $a^2 + b^2 = c^2$ by c^2, then I get

$$\left(\frac{a}{c}\right)^2 + \left(\frac{b}{c}\right)^2 = 1.$$

I let $x = a/c$ and $y = b/c$. Then the equation $a^2 + b^2 = c^2$ becomes

$$x^2 + y^2 = 1.$$

And if a, b, c are integers, then x, y will be . . . what kind of numbers?

AUDIENCE. Rational numbers.

SERGE LANG. That's right. Consequently, to find one or all the solutions of $a^2 + b^2 = c^2$ in integers is equivalent to finding all the solutions of $x^2 + y^2 = 1$ in rational numbers. Because conversely, if I have a solution (x, y) in rational numbers, then I can write each number as a fraction, with a common denominator c; and then I clear denominators and I find a solution of $a^2 + b^2 = c^2$ in integers. The problem is now to find all the solutions of $x^2 + y^2 = 1$ in rational numbers.

Do you know what the equation $x^2 + y^2 = 1$ represents? What is its graph?

AUDIENCE. A circle.

SERGE LANG. Yes, we can draw it here.

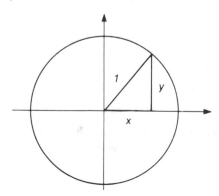

It's a circle of radius 1, and with center at the origin of the axes. We have a triangle of hypotenuse 1, and sides x, y. We can state our problem by saying that we must find all the rational points on the circle, that is all the points whose coordinates x and y are rational numbers.

Before I find *all* the solutions, I am going to write down a lot of them. I let:

$$x = \frac{1 - t^2}{1 + t^2} \quad \text{and} \quad y = \frac{2t}{1 + t^2}.$$

I write these formulas . . .

MR. A. [*Aggressive.*] But you thought of these just like that . . .

SERGE LANG. No, I did not think of them "just like that", but someone, long ago, thought of them "just like that".

MR. A. Oh yes? Really, all of a sudden?

SERGE LANG. No, of course not, he was playing with mathematics, he was looking at a lot of things, and then he realized that it gave solutions. When he realized this, he was doing mathematics and he was being a good mathematician. But once he discovered it, then the next generations use the result, and copy it. That's all I was doing, I don't claim anything else.

MR. A. Don't you think that is precisely the difficultly for someone who does not keep up with mathematics, to find these results in order to effectively do mathematics?

SERGE LANG. Where a mathematician goes fishing for these things cannot be explained. Each mathematician gets them wherever he can. Right now, I am trying to show you a complete solution of the problem. After that, I'll show you unsolved problems. You can work on them . . . you can go fishing for them by yourself, and if the fish bites and you catch a big fish, then you get a gold medal or a chocolate medal.

ANOTHER. It comes from trigonometry, no?

SERGE LANG. It comes from wherever you want. I don't have time now to show you that in greater detail. It comes from many places simultaneously.[1]

---

[1] The question where those formulas come from arises frequently, and until today, I did not know the answer. Considering the intensity of the audience's reaction, both during the talk and afterwards, I decided to look into the history of these formulas more closely. Historically the Greeks were interested in the solutions of $a^2 + b^2 = c^2$ in integers. Euclid (three centuries BC) already knew the formulas

$$a = m^2 - n^2, \quad b = 2mn, \quad c = m^2 + n^2,$$

with integers m, n. Diophantus (three centuries AD) knew how to deal with fractions, and also knew that if you divide these formulas by $m^2 + n^2$ and put $t = m/n$, then you get back the formulas which I have written above. These formulas therefore certainly did not come from "trigonometry". Diophantus was interested in finding rational solutions to equations

Now let's check that our formulas do give solutions of $x^2 + y^2 = 1$. With very little algebra, you find:

$$x^2 = \frac{1 - 2t^2 + t^4}{1 + 2t^2 + t^4}, \qquad y^2 = \frac{4t^2}{1 + 2t^2 + t^4}$$

and therefore

$$x^2 + y^2 = \frac{1 + 2t^2 + t^4}{1 + 2t^2 + t^4} = 1.$$

We have therefore found an identity, which is valid for all values of t. Suppose that I substitute for t some rational number. What do I obtain for x and y?

AUDIENCE. ???

SERGE LANG. We obtain rational numbers. We obtain them from t by additions, subtractions, multiplications, and divisions. Therefore we obtain rational numbers.

AUDIENCE. Yes.

SERGE LANG. Look at an example. Somebody—you, madam, give me some value for t.

LADY. One half.

SERGE LANG. Thank you. We put $t = 1/2$ and we compute a little:

$$x = \frac{1 - 1/4}{1 + 1/4} = \frac{3/4}{5/4} = \frac{3}{5}$$

---

just like the one we have considered, and like we shall consider later. The search for these solutions is known today under the name of diophantine problems. The equations are called diophantine equations. See [Di], especially Book VI, where Diophantus solves problems concerned with Pythagorean triangles with additional conditions, using the formulas. See the end of the conference for the converse, and also [La–Ra]. Since it may interest people to see how Diophantus expressed himself, I reproduce here the first few lines of Problem XVIII of Book VI:

To find a right triangle such that the number of its area augmented by the number of its hypotenuse forms a cube, and that the number of its perimeter is a square.

If, as in the preceding proposition, we suppose that the number of the area is one arithme, and that the number of the hypotenuse is a cubic quantity of units, minus 1 arithme, then we are led to search for a cube which, augmented by two units, is a square . . .

There are about 300 pages in this style!

and

$$y = \frac{2 \cdot 1/2}{1 + 1/4} = \frac{1}{5/4} = \frac{4}{5}.$$

Here we find the triangle 3, 4, 5. OK, 1/2 is not very big and it is natural that we found the same solution in integers that we already knew. Now, if you want to do the computation with another fraction, maybe one that is not so simple, you will find other solutions. Do you want to give me another fraction?

LADY. 2/3.

SERGE LANG. All right, let's compute quickly:

$$x = \frac{1 - 4/9}{1 + 4/9} = \frac{9 - 4}{9 + 4} = \frac{5}{13}$$

$$y = \frac{2 \cdot 2/3}{1 + 4/9} = \frac{4/3}{13/9} = \frac{12}{13}.$$

Now we got back the solution 5, 12, 13 which somebody already mentioned. It's clear that you can continue with any fraction t, or any integer t. If you substitute for instance $t = 154/295$, you will get values for x and y which are a lot bigger, and which will give solutions. By this process, you see how to obtain an infinite number of solutions. It is a theorem that one obtains all of them except one: $x = -1$ and $y = 0$ cannot be obtained by such substitutions in the formulas. But all the other solutions (x, y) in rational numbers can be obtained by this procedure, by substituting a rational value for t in the formulas

$$x = \frac{1 - t^2}{1 + t^2} \quad \text{and} \quad y = \frac{2t}{1 + t^2}.$$

Since I want to deal with another topic at greater length, I am going to skip now the proof that this gives all the solutions except one of them. Maybe there will be time to give this proof later, after the talk.

MR. A. You said that one "sees" that there is an infinite number of solutions. Who "sees" it?

SERGE LANG. If you substitute an infinite number of values of t in these formulas, you get an infinite number of values of x.

MR. A. But it's not so easy to see.

SERGE LANG. Yes, it is, but I don't want to go into details now.

MR. A. But I want to say that it cannot be seen so easily. [*Brouhaha in the audience.*]

SERGE LANG. It depends who looks at it, it depends how good your eyes are. [*Laughter.*] [2]

OK, we just considered the equation $x^2 + y^2 = 1$. Suppose we want to generalize this equation, and study others which are more complicated. What will be the next complicated type of equation that we should look at? Let's pick on somebody. Madam.

THE LADY. Replace 1 by another number.

SERGE LANG. That's a possibility. We can study $x^2 + y^2 = D$. There is a theory for that which is quite similar to the one we have just seen. Let me skip it.

AUDIENCE. Look at the equation $x^2 + y^2 + z^2 = D$.

SERGE LANG. Very good, we can increase the number of variables. This raises some very interesting questions. But I am trying to make you say what I have in mind, I am trying to make you suggest what I intend to do.

AUDIENCE. Replace the square by a cube.

SERGE LANG. There we are. For example, the equation $x^3 + y^2 = D$, obtained by putting 3 instead of 2. Let's write it in the most classical form:

$$y^2 = x^3 + D.$$

For instance, $y^2 = x^3 + 1$. Are there infinitely many solutions? Is there even a single one?

AUDIENCE. Yes. 2 and 3, because $3^2 = 2^3 + 1$.

SERGE LANG. Is there another one?

AUDIENCE. $x = 0, y = 1$; and $x = -1, y = 0$.

SERGE LANG. OK, we now have three solutions. Is there another one?

AUDIENCE. $x = 0, y = -1$.

SERGE LANG. That's right, because of the square, we can take y or $-y$. So to summarize, we have the five solutions:

$$x = 0, y = \pm 1; \qquad x = -1, y = 0; \qquad x = 2, y = \pm 3.$$

---

[2] No matter how you look at it, you will find immediately what you are looking for. For instance, we have the equation

$$x(1 + t^2) = 1 - t^2 \qquad \text{so} \qquad (1 + x)t^2 = 1 - x \qquad \text{and} \qquad t^2 = \frac{1 - x}{1 + x}.$$

So to each value of x there corresponds a value of t or $-t$, and at most two values of t give the same value of x.

One can also notice that if t increases from 0 to 1 then $1 - t^2$ decreases, while $1 + t^2$ increases, so $x = (1 - t^2)/(1 + t^2)$ decreases from 1 to 0. In particular, different values of t give different values of x.

Are there other solutions? Who says yes? Raise your hand. Who says no? Who keeps a prudent silence?

It's not at all trivial. The difficulty to find solutions for this equation, and other similar ones, is much greater than for the equation $x^2 + y^2 = 1$. It is a theorem that there is no other solution. It's out of the question to prove it here.

Now, who knows about graphs? Do you know how to draw a graph? Who does not know? Raise your hand so I can see. [*A few hands go up.*] OK, I shall explain briefly what a graph is.

Suppose I have here on this axis values of x, and on the other axis values of y, and each x is a real number. For any real number x, I cube it, I add 1, and then I find two values for y:

$$y = \sqrt{x^3+1} \quad \text{and} \quad y = -\sqrt{x^3+1}.$$

For example:

if $x = 1$, then $y = \pm\sqrt{2}$;
if $x = 2$, then $y = \pm 3$;
if $x = 3$, then $y = \pm\sqrt{28}$;
if $x = -1$, then $y = 0$.

If x is negative and smaller than -1, then $x^3 + 1$ is negative, and there won't be a corresponding value for y. On the opposite side, if x grows indefinitely, then y grows also. To each x there correspond values y and $-y$, as on the following figure.

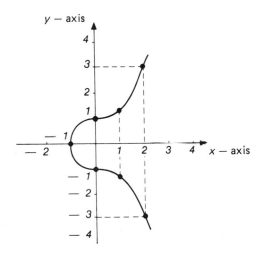

We can generalize our equation as you wanted to do earlier for $x^2 + y^2$, by considering

$$y^2 = x^3 + D, \quad \text{where } D \text{ is positive or negative}.$$

We also want to consider the equation $y^2 = x^3 + x$, or $y^2 = x^3 + ax$, which has considerable historical interest. For example, the Greeks and the Arabs had raised the following question. What are the rational numbers A such that A is the area of a right triangle, with integral sides a, b just like those we considered at the beginning. One can show that A is such a number if and only if the equation

$$y^2 = x^3 - A^2 x$$

has infinitely many rational solutions.[3]

So finally, let's consider the equation

$$y^2 = x^3 + ax + b,$$

which covers all these cases. When we dealt with $y^2 = x^3 + b$ or $y^2 = x^3 + ax$, we assumed that $b \neq 0$ and $a \neq 0$, otherwise the equations are too degenerate. Similarly, for the general equation, we assume that $4a^3 + 27b^2 \neq 0$, to guarantee the appropriate non-degeneracy. For our purposes, you don't need to pay any more attention to such a technicality.

The graph of the general equation $y^2 - x^3 + ax + b$ is going to look like this, with a branch tending to infinity, and sometimes an oval.

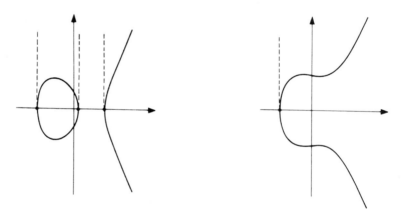

---

[3] The area of a right triangle whose sides are a, b and hypotenuse c is given by the formula

$$A = ab/2.$$

Hence we find

$$c^2 + 4A = a^2 + b^2 + 2ab = (a + b)^2$$

$$c^2 - 4A = a^2 + b^2 - 2ab = (a - b)^2.$$

It follows that a rational number A is the area of a right triangle if and only if one can solve simultaneously the equations

$$u^2 + 4Av^2 = w^2$$

$$u^2 - 4Av^2 = z^2$$

Using this graph, we can define an addition for points. Take two points P and Q on the curve. We define the sum of these two points in the following way. The straight line passing through P and Q intersects the curve in a third point. We reflect this point over the x-axis, and we find a new point which we denote by $P + Q$ as on the figure.

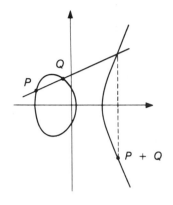

a) case where there is an oval   b) case where there is no oval

A UNIVERSITY STUDENT. But it does not always happen that the line through two points intersects the curve in a third point.

SERGE LANG. Oh yes? Can you give me an example?

A STUDENT. Well, yes, if the line is vertical.

SERGE LANG. Excellent remark. She is right, because if Q is the reflexion of P over the x-axis, then the vertical line will not cut the curve in any other point. We shall come back to this special situation in a moment. But this is essentially the only possible example of this phenomenon. Before looking at this special case, let's go back to the definition of the sum of two points.

I have used the symbol $+$. You have the right to expect certain properties, otherwise I should not have used the symbol $+$. What are those properties?

AUDIENCE. ???

---

in rational numbers (u, v, w, z). In a recent article, J. Tunnell [Tu] took up this theme and remarked that if one makes a projection from the point $(1, 0, 1, 1)$ onto the plane $z = 0$, then one obtains a correspondence between the curve defined by these simultaneous equations, and a plane curve, which itself can be put in the form

$$y^2 = x^3 - A^2 x,$$

which is precisely of the type we are now considering. Tunnell gives criteria for the existence of an infinite number of solutions depending on recent, and quite difficult mathematical theories.

SERGE LANG. You know the symbol + from ordinary addition of numbers. I have just defined an addition of points. Which properties does addition of numbers have?

SOMEONE IN THE AUDIENCE. It's a group law.

SERGE LANG. Don't use such fancy language.

SOMEONE ELSE. The order of the terms can be reversed.

SERGE LANG. Indeed, that's the first property. We must have

$$P + Q = Q + P.$$

Which is true. To compute $Q + P$, I use the same straight line, so I find the same point of intersection, so the same sum $Q + P = P + Q$. What other properties can you expect?

SOMEONE IN THE AUDIENCE. Associativity.

SERGE LANG. You, it's clear that you know too much. [*Laughter.*] Let others speak too. For instance, the lady, there.

LADY. Associativity.

SERGE LANG. Yes, that's right. What does it mean? If I take the sum of three points, I could take it in two possible ways:

$$P + (Q + R) \quad \text{and} \quad (P + Q) + R.$$

Associativity means that these two expressions are equal, therefore we have

$$P + (Q + R) = (P + Q) + R.$$

It's obvious that $P + Q = Q + P$, but if you try to prove associativity, you won't find it so easy. If you try by brute force, you won't succeed. But it's true.

What other properties do you expect?

A HIGH SCHOOL STUDENT. A neutral element?

SERGE LANG. That's it. So what will be the neutral, or zero element? It means an element such that

$$P + \text{neutral element} = P.$$

Is there one?

SOMEONE. The point over there.

SERGE LANG. No. This requires some imagination. Ah [*Laughing*] the gentleman over there does like this [*pointing upward.*] Are you a mathematician?

GENT. No, but I have been one. [*Laughter.*]

SERGE LANG. We are forced to invent this neutral element. Let's redraw the figure.

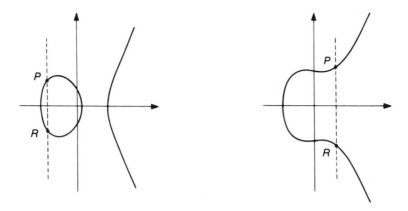

You are given a point P. What do I have to find? I must find something which is such that, when I take the straight line between P and this something, the line cuts the curve in a point whose reflection over the x-axis is P itself. The reflection of P is denoted by R in the figure, and the line passing through P and R is the vertical line. Consequently, if there is a point O such that $P + O = P$, this point cannot lie anywhere in the plane, because it must be on the curve and on the vertical line. So what do we do? We invent this point. We call it zero, and denote it by O. We say that O is at infinity. All the vertical lines tend toward infinity, going up or down. We make the convention that all these points at infinity are all the same point. We define a single point at infinity, which we view as the intersection of all vertical lines. It is a convention we accept that the straight vertical line passing through P also passes through P and O, and if this line cuts the curve at R, then $P + R = O$. Then what should we call R?

AUDIENCE. Minus P.

SERGE LANG. Yes, very good, because one has the condition

$$P + (-P) = O.$$

That's the convention we adopt.

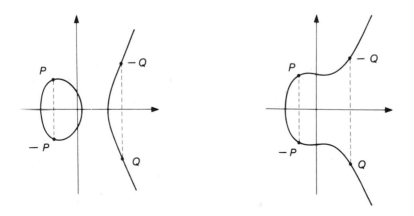

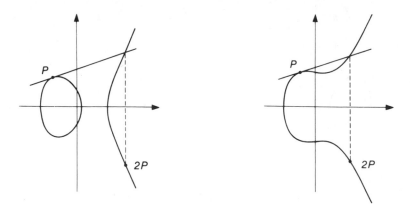

And if I want to find $P + P$, what do I do?

AUDIENCE. Take the tangent.

SERGE LANG. That's right, perfect. The tangent to the curve at P cuts the curve in a point, which we reflect to obtain $P + P$, which I also denote by $2P$. Suppose I want to find $3P$. What do I do? I take the sum $2P + P$, always following the same process: I draw the line between P and $2P$, I reflect the point of intersection of this line with the curve, and I find $3P$. Same thing for

$$4P = 3P + P, \qquad 5P = 4P + P, \qquad \text{and so on}.$$

Now a little question. Where are all the points P such that $2P = O$? Use your imagination. Where are they? You. [*Pointing to someone.*]

SOMEONE. I don't see.

SERGE LANG. You saw how we find $2P$. We draw the tangent, we look at where the tangent cuts the curve, we reflect, and we get $2P$. Now I want $2P$ to be at infinity.

GENTLEMAN. On the horizontal line.

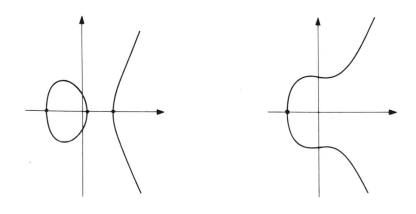

SERGE LANG. That's right, the points P such that $2P = O$ will be the points whose tangent is vertical, and therefore the points on the curve which lie on the horizontal line, the x-axis. There will be three such points if there is an oval. If there is no oval, then there is only one such point. Plus O itself, of course.

Suppose we have found a point $P = (x, y)$ which is rational, that is whose coordinates (x, y) are rational numbers.

Then in general, I can find other rational points: the multiples $2P$, $3P$, $4P$, etc. will also be rational. One can see this because one can give a formula for the addition of two points.

Let's look at three points on the curve $y^2 = x^3 + ax + b$:

$$P_1 = (x_1, y_1), \qquad P_2 = (x_2, y_2), \qquad P_3 = (x_3, y_3)$$

and suppose that $P_3 = P_1 + P_2$. How do you compute the coordinates x_3, y_3 from the coordinates x_1, x_2, y_1, y_2? There is a formula:

$$x_3 = -x_1 - x_2 + \left[\frac{y_2 - y_1}{x_2 - x_1} \right]^2.$$

Of course if $x_1 = x_2$ then the formula does not make sense. In this case, if $P = (x, y)$, then we compute $2P$ by the formula

$$x_3 = -2x + \left[\frac{3x^2 + a}{2y} \right]^2.$$

[*Six persons leave the audience at this point.*]

These formulas, again, nobody can find them just like that. They lie much deeper than those which give rational points on the circle. One needs serious ideas, general ideas, in order to arrive at the notion of straight line between two points intersecting the curve in a third point. But if you follow this procedure, and if algebra does not give you any trouble, then you will be able to derive these formulas with about a page of computations.

Let's apply these formulas to find rational points. We pick a concrete example, for instance the equation

$$y^2 = x^3 - 2.$$

There is a first solution, $x = 3$ and $y = 5$. Let's call this solution P. Then Mr. Brette (who organized the conference at the Palais de la Découverte), was kind enough to do the computations with a computer, in order to find multiples of P. The solution $2P$ has coordinates $2P = (x_2, y_2)$ where

$$x_2 = \frac{129}{100} \quad \text{and} \quad y_2 = \frac{-383}{1000}.$$

He substituted $x = 3$ and $y = 5$ in the formula for $2P$. Then he went on to find the following table.

The curve $y^2 = x^3 - 2$
Multiples $nP = (x_n, y_n)$ of the point $(3,5)$

n	x_n	length
1	3 1	1
2	129 100	3
3	164323 29241	6
4	2340922881 58675600	10
5	307326105747363 160280942564521	15
6	7948453616231848880769 51312731007360614490 0	21
7	4968031750452922778611893792 3 3458519104702616679044719441	29
8	300370887246304508033820355385035 05921 301068398289876307178684299377991 8400	38
9	182386897568483763089689099250940650872600619203 127572396335305049740547038646345741798859364401	48
10	2916788280313095843339723491701940084224073562766495053324 9 133299362853639218191065074976813043197323638162648320250 0	59
11	824172661141552804187727190037944704511772520763880755114120154630088 03 91802056604733629263993982598037348195816860458964963953593942680660 1	71

For reasons of space, the values of y_n are omitted. One can compute them by the formula

$$-y_3 = \left[\frac{y_2 - y_1}{x_2 - x_1} \right] (x_3 - x_1) + y_1.$$

If you look at the numerator of x_n in this table, you see that the numerators of these fractions increase very regularly. In fact, the way they increase has been one of the fundamental problems in this field of mathematics, diophantine equations. I have shown you the simplest example, after the equation of the circle.

The problem is to find all solutions, in integers or rational numbers, for equations of this type. It's extremely hard. There is no known procedure today to determine all the solutions. For the special equation

$y^2 = x^3 - 2$, I was able to write down one solution by inspection. But if I give you an equation of this type, there is no systematic method which allows you to find a first solutions, by an effective process. It's one of the great problems that mathematicians face: find a first solution by an effective process. But if I give you a first solution, then you can find others by applying the formulas.

Two cases can happen. The first is like the case with $2P = O$, but when, instead of $2P$, we might have $3P = O$, or $4P = O$, or $5P = O$. In general, if P is a point on the curve such that

$$nP = O$$

with some positive integer n, then we say that P is a point of finite order, or of order n. One question is to find out if there exist many rational points of finite order. One of the greatest discoveries of modern mathematics is due to Mazur, just three or four years ago [Maz], that if P is a rational point of order n, then n is at most 10, or $n = 12$. Furthermore, there are at most 16 rational points of finite order.[4]

The second case is when you construct $2P, 3P, 4P, \ldots$ you find a new point each time, like in the table a minute ago. You find points whose size grows regularly.

I am now going to show you, in the remaining few minutes, what are some of the theorems and conjectures concerning such equations and their solutions.

In 1922, Mordell [Mo] proved a conjecture of Poincaré [Poi] that one can always find a finite number of rational points

$$P_1, P_2, P_3, \ldots, P_r$$

such that any rational point P can be written as a sum of these points; this means that there are integers $n_1, n_2, \ldots, n_r$ depending on P, such that P can be written as a sum,

$$P = n_1 P_1 + n_2 P_2 + \cdots + n_r P_r.$$

Addition is, of course, addition on the curve as I have defined it.

[*Someone raises his hand.*]

SERGE LANG. Yes?

A HIGH SCHOOL STUDENT. What's "r"?

SERGE LANG. That's a very good question. There might be relations between the points $P_1, \ldots, P_r$. For example, one of them could be of finite

---

[4] Mazur's methods are among the most advanced of contemporary mathematics, and depend on what is called algebraic geometry and the theory of modular curves.

order. One can prove that we can always choose the points $P_1, \ldots, P_r$ such that any rational point can be expressed as a sum

$$P = n_1 P_1 + n_2 P_2 + \cdots + n_r P_r + Q,$$

with integers $n_1, \ldots, n_r$ which are uniquely determined by P; and Q is a point of finite order. This means that there are no relations among the points $P_1, \ldots, P_r$. If I choose r like that, then r is the maximum number of points among which there are no relations. Since Poincaré, r is called the rank of the curve. The problem is to determine r and to find the points $P_1, \ldots, P_r$.

Nobody knows how to do it in general. In special cases, one has methods which give a solution to the problem. Here is a table of Cassels for curves $y^2 = x^3 - D$, where D is an integer between -50 and $+50$. In each case, the rank is 0, 1, or 2. Cassels give the points P_1, P_2 as they arise [Ca]. [*You will find the table in an appendix.*]

For the general case, there exist very deep conjectures; one of them is due to Birch and Swinnerton-Dyer, two English mathematicians [B–SD]; it gives the rank in terms of very complicated objects associated with the equation. I cannot enter here into these considerations. But you can see how little we know, since nobody today knows an example when the rank is large, nor even an example when the rank is bigger than 10 (I think, it may be 12). Still, mathematicians conjecture that there are cases when the rank is arbitrarily large. Anybody can think about this problem: find a curve with an equation

$$y^2 = x^3 + D,$$

with D an integer, whose rank is bigger than 15, or 20, or 100, or arbitrarily large. We believe that such curves exist, but it's a great challenge to find them.

Recently, Goldfeld formulated the question somewhat differently [Go]. He considers curves

$$Dy^2 = x^3 + ax + b,$$

where a, b are fixed and D varies. Let's say D is an integer, $D = 1, 2, 3, 4$, etc. How does the rank behave for these values of D? For instance, how many integers D are there less than or equal to a number X for which the rank is 0, so for which there won't be any rational point except possibly a point of finite order? How many $D \leq X$ are there for which the curve has rank 1? How many $D \leq X$ are there for which the curve has rank 2? And so on. Goldfeld suggested that one should find a fairly regular behavior for rank 0 or 1; in fact he expects that the density of each should be one half for rank 0 and one half for rank 1. This means that approximately half the curves should have rank 0, and half of them should have rank 1, with perturbations which depend on much more complicated invariants of the curve. And there should be relatively few values of D for which the rank is bigger than 1.

It is a fundamental problem to give a quantitative answer to questions like that, and similar questions for curves like

$$y^2 = x^3 + D,$$

with D variable, or for the general family of curves $y^2 = x^3 + ax + b$ with a and b variable: for which values of a, b do we get rank 0, 1, 2, 3, 4, or any given integer as the rank. Since we don't even know whether there exist such curves with rank bigger than 10, we are far from knowing the answer, except possibly conjecturally.

Euh . . . that's a lot of algebra. I hope it wasn't too much. I just wanted to try, and see if I could make you understand this kind of problem that mathematicians raise. But I have been talking for an hour, so I'll stop and we'll see if there are any questions and if you have gotten anything out of all this.

The questions

SOMEONE. What is it good for?

SERGE LANG. I already gave the answer last year: it's good to give chills in the spine to a certain number of people, me included.[5] I don't know what else it is good for, and I don't care. But I speak for myself only. Like von Neumann said, one never knows whether someone is going to find another use for it. I was just trying to show you the kind of problem that excites us, or that excites me.

A HIGH SCHOOL STUDENT. This kind of problem is analogous to someone doing research in physics or electronics. They do experiments, but they don't know what they will find. It's like penicillin, for instance.

SERGE LANG. There is no universal answer, but your comment is very valid.

A GENTLEMAN. There is a question which interests me very much: it's the hyperdimensions of space. I hear that Lobatchevski found up to thirty-two dimensions. Do you believe that's a limit, or are there more?

SERGE LANG. I don't know what you mean by hyperdimensions.

GENTLEMAN. You don't know what hyperdimensions mean? Do you believe there are only three dimensions in space?

SERGE LANG. If you put the question that way [*Laughter*] then I can give, if not an answer, then at least an analysis of the question. You asked me: "Do you believe there are only three dimensions in space?" What do you mean by "space"? If by space you mean "that" [*Serge Lang shows the room*] then by definition there are only three dimensions. If you want more dimensions, then you accept to give the word "dimension" a more

---

[5] Not to speak of Diophantus . . .

general meaning, which is anyway the one which has been accepted long ago. Every time you can associate a number with a notion, you have a dimension, no matter what kind of notion you start from; in physics, mechanics, economics, or anything else. In mechanics, besides the three spatial dimensions, you can have speed, acceleration, curvature, etc. In economics, take for example the big businesses, oil companies, the sugar companies, steel, agriculture, etc. and their gross profits in 1981. For each company you get a number, and therefore a dimension; and in addition, of course, the number 1981 associated with time. Then you can have hundreds of dimensions like that.

By the way, if you look in the Encyclopoedia of Diderot, under "dimension", you will see that d'Alembert wrote the comments, and here is what he wrote:

> This way of considering quantities of more than three dimensions is just as right as the other; because algebraic letters can be seen as representing numbers, rational or not. I have said above that it was not possible to conceive more than three dimensions. A clever gentleman friend of mine believes that one could nevertheless view duration as a fourth dimension, and that the product of time by solidity would in some way be a product of four dimensions. This idea can be challenged, but it has, it seems to me, some merit, were it only that of being new. [Did]

Naturally, a friend of his, that's him, but he is being careful. He understood that the notion of dimension should not be restricted to space, but could be associated with any situation when you can associate a number. Time is only one example.

The rank of curves which we discussed before is another example. We can say that if a curve has rank r, then the rational points generate a space of dimension r.

SOMEONE. Does it help you in your theories to be able to use computers to find solutions, may be not all solutions, but some of them?

SERGE LANG. Yes, definitely. The Birch and Swinnerton-Dyer conjectures were based on experimental data from computers, as well as intuition and theoretical results. Historically, the rate of growth of the length of the multiples of one point could have been discovered by computers. More precisely, if you have a rational point $P = (x, y)$ on the curve, write $x = c/d$ where c is the numerator and d the denominator. Write

$$nP = (x_n, y_n) \qquad \text{with} \qquad x_n = c_n/d_n.$$

Then how fast does c_n grow? It is a theorem due to Néron that the length of c_n grows approximately like n^2. In the table of multiples of P, you can see this growth illustrated for $n \leq 11$. To make more precise what we mean by "approximately", I need a more elaborate mathematical language. I would have to say that it is a quadratic function, up to a bounded function. I don't want to go into this now. One can write down a more precise formula for the length, but it's much more difficult.[6] Here I merely stated an approximate behavior for the length.

SOMEONE. Is there some relation between the addition of points that you showed us, and the question of strange attractors?

SERGE LANG. Strange attractors in what, physics?

THE PERSON. Yes, systems of iteration which give certain kinds of curves.

SERGE LANG. Are you a physicist?

THE PERSON. Yes.

SERGE LANG. I don't know your physics and you don't know my elliptic curves. Maybe it's time we should get to know each other. I don't know an answer to your question, I don't know much physics. But it's possible. [*To the audience:*] Do you see, what's happening right now? I wrote certain formulas which struck a chord in the gentleman's mind. They suggested something to a physicist, by free association of ideas. That's how one does research. Two things can happen. Either nothing comes of it, or the gentleman will pursue the idea, which perhaps will give new relations between certain physical theories and the theory of so-called elliptic curves—of cubic equations. Maybe we'll know next year. The physicist might give a conference on those relations. That's what research is. But right now, I don't know the answer.

A GENTLEMAN. Can you tell us something of Fermat's great theorem?

SERGE LANG. Fermat's conjecture?

GENTLEMAN. Yes.

SERGE LANG. One can generalize the equation we looked at, for example, we can consider $x^3 + y^3 = 1$, or more generally

$$x^n + y^n = 1$$

where n is an arbitrary positive integer. What happens when n goes from 3 to 4?

SOMEONE. There are no solutions!

---

[6] Let us write x as a fraction, $x = c/d$ where c is the numerator and d the denominator. Define the height of the point to be

$$h(P) = h[x(P)] = \text{maximum of } \log |c|, \log |d|.$$

Néron's theorem states in particular that $h(nP) = q(P)n^2 + O(1)$, where $q(P)$ is a number depending of P, and $O(1)$ is a term bounded independently of n. The number $q(P)$ is called the quadratic form of Néron–Tate, because Tate gave a very simple proof for its existence. Mathematicians raise many questions about this number $q(P)$, for example whether it is a rational number or not. People believe it is not, unless P is of finite order. One can define a distance between two points P and Q by letting the square of this distance be $q(P - Q)$. The study of this distance constitutes one of the fundamental problems of the theory.

SERGE LANG. Sir, you know too much, it's cheating. Don't butt in. Besides, there are solutions:

$$x = 1, y = 0 \quad \text{and} \quad x = 0, y = 1.$$

[*Laughter.*] Are there others than those with $x = 0$ or $y = 0$? Who says yes? Who says no? Who does not know the answer? [*There are still some people who did not raise their hands.*] Who thinks that the answer is known? [*Laughter.*] Who thinks that the answer is not known? [*Several hands go up.*] Who knows that the answer is not known? [*Laughter.*]

In fact, the answer is not known. One knows the answer for a large number of values of n, but not in general. That's Fermat's problem:

Are there solutions of $x^n + y^n = 1$ in rational numbers, other than with $x = 0$ or $y = 0$, when n is an integer > 2?

The answer is not known in general. One believes that the answer is no.

A HIGH SCHOOL STUDENT. Do people hope to know the answer some day?

SOMEONE ELSE. But Fermat said that he knew the answer!

SERGE LANG. Yes, Fermat said that[7] but one still does not know it. As for the question if one hopes to know the answer some day, what does it mean?

THE STUDENT. Does humanity hope to know the answer? Is it provable, or has it been shown to be unprovable?

SERGE LANG. No, it's an act of faith that it's provable. Mathematicians—euh, to be careful, all those I know—[*Laughter*] believe that it's provable. I think that if you raise an intelligent mathematical problem, there is an answer which will be found, some day.[8] That means, it suffices to think about the problem, and somebody will find the solution. Problems which are not solvable, that is, for which one can prove that they are not provable one way or another, are pathological cases, and I don't care about them. They don't occur when one "does mathematics". You have to look for them specifically.

SOMEONE. What's the definition of an intelligent problem?

SERGE LANG. No definition. [*Laughter.*]

The problems that you will meet, like that, it's an act of faith by mathematicians that you can try to solve them, and that you will succeed.

---

[7] More precisely, Fermat used to write comments in the margin of Diophantus' collected works. Next to the problem where Diophantus gives solutions of Pythagoras' equation $a^2 + b^2 = c^2$, Fermat wrote that he had a "marvelous" proof of the fact that for higher degree, there are no other solutions besides the trivial solutions, but the margin was too small to write down his proof.

[8] My use of the word "intelligent" is obviously idiotic, and the following sentences are deficient in that they don't take into account properly the choice which everyone makes concerning the subject of one's research.

That's all. One does not even think of the possibility that they are perhaps not provable. And if you think too much about that, then maybe you will do something else, but you won't do this kind of mathematics. It will prevent you from thinking.

But watch out! There are some problems which are somewhere in between, for example what is called the continuum hypothesis. It is the only counterexample that I can think of right now.

QUESTION. What is the continuum hypothesis?

AUDIENCE. Cantor . . .

SERGE LANG. Yes, let's talk a little about the continuum hypothesis. Last year, somebody got chills in the spine just to know whether the real numbers are denumerable. Take all the real numbers, the numbers on the number line, or in other words, all the infinite decimals, like

$$212.35420967185\ldots$$

You also have the positive integers 1, 2, 3, 4, . . . One says that a set is denumerable if you can make a list of all the elements of the set, with a first, a second, a third, and so on, so that you catch all the elements of the set, so that none is left out. Somebody last year asked me to prove that the real numbers are not denumerable, and I gave the proof.

Mathematicians, or Cantor, raised the following question. Between denumerable sets, those that you can enumerate like the integers, and the real numbers, are there sets whose cardinality is in-between; that is, sets which have more elements than the denumerable sets—so that you cannot enumerate them—but which have fewer elements than the real numbers? What does it mean, "fewer"? It means that you cannot establish a one to one correspondence between the real numbers and the elements of this set. The continuum hypothesis was that there does not exist any such sets, non-denumerable, but with "fewer" elements than the real numbers.

Considering the way we write the real numbers, as infinite decimals, they seemed so close to the rational numbers (which are denumerable), that it seemed reasonable to think that there was no set of intermediate cardinality.

SOMEBODY. Maybe someone is trying to find the answer?

SERGE LANG. Of course, that's why I said that it was a counterexample to the statement I made. There is no doubt that the question is intelligent. And the solution was found by somebody who did not get caught by the way the question was phrased. It's Paul Cohen.

QUESTION. What century?

SERGE LANG. Recent, about fifteen years ago. And the answer is that the question is meaningless. One can prove neither that there exists such a set, nor that there does not exist such a set. The answer is that, given the mathematical system with which we work today, which is sufficient for all our needs except this one, if you add as an axiom the positive answer to the continuum hypothesis, then you still have a consistent system, the system will still be valid. And if you add as an axiom the negative answer to the continuum hypothesis, then again the system will also be consistent.

AUDIENCE. It's independent of the axioms you already have.

SERGE LANG. That's right. What I mean is that the questions was badly posed. It means that when you speak of "sets", you don't know what you are talking about. The ambiguity lies in the intuitive notion you have of a set. Everybody has some intuition of sets: a set is a . . . bunch of things. [*Laughter.*] To say a bunch of things, it's OK if you speak of all the real numbers; it's OK if you speak of all the rational numbers; it's OK if you speak of all the points on a curve; but if you speak of all sets simultaneously, of all the sets contained in the real numbers, then it's not OK, it does not work any more. That's what Paul Cohen's answer means: our notion of set is too vague for the continuum hypothesis to have a positive or negative answer. There remains that many mathematicians feel the need of an axiom which is psychologically satisfactory, and which would imply either the continuum hypothesis or its negation. This side of mathematics is interesting to some people. It does not really interest me personally. But I have to admit that it was worth seeing: a question which nobody thought could have an answer other than yes or no; and the guy who answered: you are all wet, there is no possible answer.

THE HIGH SCHOOL STUDENT. Is it possible that Fermat's conjecture is of this type?

SERGE LANG. What do you want me to answer? From my point of view, it's obvious what I am going to answer. It's not me that's going to say that it could be of the same type. No way.

Besides, there is an argument . . . [*hesitates*] if you succeeded in proving that Fermat's problem is unsolvable, then ipso facto you would have shown that the conjecture is true. Because if there was a counterexample, then with some big computer, some day someone would pull out the counterexample. But I hate this type of argument, and as far as I am concerned, I regard it as the normal state of affairs that some day, somebody will prove Fermat's theorem, or will prove that it is false.

QUESTION. And you personally, do you believe it is true or false?

SERGE LANG. [*Hesitates*] Well, it's true. There is no other solution besides $x = 0$ or $y = 0$. For the following reasons. We begin to understand the theory of such equations from a general point of view. There is a general conjecture of Mordell which I am going to describe.

Take an equation, for instance

$$y^3 + x^2y^7 - 312y^{14} + 2xy^8 - 18y^{23} + 913xy + 3 = 0.$$

This is what one calls a general diophantine equation. We ask in general: are there infinitely many solutions of this equation in rational numbers x, y? We have already seen two types of examples when there exist such solutions. In the first example, we could express x as a quotient of two polynomials in a variable t, and y similarly, so that the equation was satisfied as an indentity of t. This is precisely what happened we used the formulas

$$x = \frac{1 - t^2}{1 + t^2} \quad \text{and} \quad y = \frac{2t}{1 + t^2},$$

and found that $x^2 + y^2 = 1$, an identity in t. Clearly (despite somebody's objections), you will get infinitely many solutions. That's one of the possibilities.

The other possibility is that you can get solutions of the equation from a cubic, with formulas

$$x = R(t, u) \quad \text{and} \quad y = S(t, u),$$

where t, u satisfy an equation $t^2 = u^3 + au + b$ having infinitely many solutions; and R, S are quotients of polynomials, with rational coefficients.

The first possibility is called genus 0, and the second is called genus 1.

Mordell's conjecture says this. Let $f(x, y) = 0$ be an equation, where f is a polynomial with integer coefficients. If you cannot reduce this equation to the case of genus 0 or genus 1 by formulas like the above, then the equation has only a finite number of rational solutions. That's the conjecture.

In a family of equations like Fermat's, with n variable, there should be very few solutions. One can even prove that for $n \geq 4$, the equation $x^n + y^n = 1$ cannot be reduced to genus 0 or genus 1. According to Mordell's conjecture, Fermat's equation should have only a finite number of solutions in rational numbers x and y. Some people have done computations going quite far, maybe up to $n = 1,000,000$, and one knows that up to that point, there are no solutions other than the obvious ones with $x - 0$ or $y = 0$. And if what we feel is true, then there should not be any others for even bigger values of n, because such families should behave in a regular way. If one has not found solutions at the beginning, for n small, then there should not be any later, when n is large. That's the general intuition which directs us when we work on diophantine equations. Well, OK, it's a working hypothesis. One is always ready to backtrack if somebody shows that it's wrong. That's how mathematicians work: we make working hypotheses, we try to prove something, but we are always ready to accept any evidence that we are wrong, and that we have to start over again.

[*Someone raise his hand.*] And the computers, can't you do anything with them?

SERGE LANG. Oh, the computer, it has been used many times. It is with computers that people have shown that there were no solutions up to n approximately 1,000,000.

QUESTION. Sir, I have a question—there are problems which were solved first with restrictive hypotheses, and then better mathematicians could eliminate these hypotheses. But still, the first proofs used these hypotheses. Why?

SERGE LANG. When you try to solve a problem, you try first to solve special cases, and then try more general cases. The first ideas that you have might work only in the special cases. Maybe other ideas are needed in the more general cases. Who knows when these new ideas will come? Or even if they will come to one person and not another? Somebody publishes a first paper, then someone else relies on these first results, and obtains further results, publishing a second paper, but with some new ideas; and so forth until the general problem is solved. That's how one works. It does not mean that the mathematician who succeeds in eliminating the restrictive hypotheses is "better" than the other. Quite the contrary, the first mathematician might have shown much more imagination, and might have opened up a whole domain of research where nobody understood anything before. It may be that this first contribution will be admired much more than the following ones which, perhaps, merely developed the first one's program.

QUESTION. Let me change the subject a little. At the beginning of your talk, you alluded to the teaching of mathematics in France . . .

SERGE LANG. Everywhere, in the whole world.

QUESTION. The subject is of current interest. How do you see things in this direction? There seems to be a general problem.

SERGE LANG. How do I see things? I don't understand the question. It's too general.

A HIGH SCHOOL STUDENT. Do you think that mathematics should be taught like that, just for the beauty of it and not for applications to physics, or that at least until the end of high school, they should be turned towards physics, toward applications?

SERGE LANG. The way you phrase your question is too . . . exclusive. One does not prevent the other. It's obvious that the negation of one extreme does not imply an extreme on the opposite side. Do what . . . what comes naturally. Of course, there should be applications when teaching mathematics. But from time to time, you must also be able to say: OK, let's look at $x^2 + y^2 = 1$ and let's find all the rational solutions. Some will like it, some won't like it, but I know it's the sort of thing students like. I know it because I have talked about this problem to 15 and 16 year old kids several times, and they like it. They thought it was interesting. At the beginning of the talk, they know one solution, maybe some student knows another, maybe still another, but usually nobody knows any more. And then, after five minutes, we succeed in giving infinitely many! Listen, you would have to be really insensitive not to react positively. [*Laughter.*] Well, OK, this does not mean that you should not also do applications.

QUESTION. When you are at Yale, do you have the same approach to teaching?

SERGE LANG. Same as what? Here? Yes, of course, like this. [*Serge Lang points to someone. Laughter.*] Naturally! How else do you want me to do it? Today, I was caught a little short, I picked a topic . . . I wanted to see just how far I could go in doing mathematics with you. It was hard. Because I needed algebraic formulas, it's dangerous for a Saturday afternoon audience. [*Laughter.*] Don't think I was not conscious of the difficulty. [*Allusion to the six persons who left after the first formulas.*] I just wanted to see how it would go. It did not go so badly.

AUDIENCE. No, no!

SERGE LANG. Were it only, for instance, him, or him [*pointing*], or the physicist over there. It's clear that they got something out of it, each one something different. Even if there had been only these three, it was worth it, and there were many others. Even if some of you are hung up on the formulas, if you are still sitting here, nobody is forcing you.[9]

QUESTION. Is there any hope to solve the great mathematical problems which have not yet been solved?

SERGE LANG. That's what mathematicians do, research. They hope to solve the problems which have not yet been solved. If they did not have that hope, they would not be, by definition, mathematicians doing research.

QUESTION. But you also find problems?

SERGE LANG. Yes, of course. To find the problem of which one is going to work, on which I am going to concentrate, is at least as important as solving it. To do mathematics, it is also to find problems, to make conjectures. For example, following Goldfeld, I raise the problem of finding the asymptotic behavior of the rank in a family of curves

$$y^2 = x^3 + D,$$

for example when D varies, for a given rank > 1. The density should be 0, but maybe there is an asymptotic behavior, so bounded from below, which would be much stronger than simply finding curves with arbitrarily high rank.

QUESTION. Perhaps in teaching mathematics, at least at the beginning, there is too much emphasis on solving problems instead of showing

---

[9] At the beginning of the talk, the room was about full, with about 200 persons. During the question period, about half remain.

how to pose problems. That's why I come back on what you have said; some people have suggested modelization, or similar things in applied mathematics. It's very positive: ask questions related to simple problems, before starting to solve them. Perhaps that is where the teaching of mathematics is deficient.

SERGE LANG. There is no single place where it is deficient, there are always several. If you show me the books, I'll tell you concretely from the books. I cannot give a general recipe, just like that. I like to deal with concrete instances. I'll show you in the book what I think is deficient in itself. There are always many deficiencies depending on the teacher, depending on the class, depending on a whole lot of circumstances, internal and external. No matter what I said, I did not mean that there was a single reason or a single condition which caused the deficiencies.

QUESTION. Perhaps it would be useful to enumerate these deficiencies.

SERGE LANG. Maybe, but after that, one would have to . . . Listen, I wrote up last year's talk. It's right here. This is what I have to say. I said it, I did it. I am doing it again this year, the conference will again be published. You see how I express myself, how I do mathematics. It's serious business. But it does not mean that someone else should do exactly as I do, just this way. Different people react differently. Do as you like, after all. My point of view is never exclusive. I speak only for myself, I don't like generalizations.

A HIGH SCHOOL STUDENT. I am a high school student, and there is something which I object to in the teaching of mathematics.

SERGE LANG. What year?

STUDENT. 11th grade. And since I was very small, I was shown proofs, but I was not shown, to use your analogy with music, the beauty in them. There was no taste to what was done in school. When one does music, then one gets into the beauty of the music, not just its rythm, or the theory of music . . .

SERGE LANG. In any case, the beautiful proofs, they are not in the curriculum. There is a whole lot of beautiful ones, and usually they are omitted. But anyway, did you like what I did today, these structures, the diophantine equations?

STUDENT. Yes.

SERGE LANG. Are you into computers?

STUDENT. Yes.

SERGE LANG. Where, here?

STUDENT. No, in my school, in the suburbs. But if you want, I think a priori that people who would be exposed just like that to what you did today, they might not see the beauty in it, anyway not everybody.

SERGE LANG. Of course, in an aesthetic situation, there are some who see it right away, there are some who see it later, and there are some who never see it. This is typical of an aesthetic situation. I don't ask everybody to find what I did beautiful. But still, the formula we had,

$$x_3 = -x_2 - x_1 + \left[\frac{y_2 - y_1}{x_2 - x_1} \right]^2,$$

it's a little complicated, but the fact that it can give you infinitely many solutions for the equation, I find it fascinating. I don't know what you think, but you asked enough questions to show that you are reacting positively.

THE PHYSICIST. It seems that in French schools, the main reason for the heavy-handedness and lack of understanding is that, behind the whole program, one tries to show, even to very young children, a logical construction which is completely irrefutable. Whether it is in physics or mathematics, a teacher can never allow himself to assert something without giving a clear proof for it.

SERGE LANG. I entirely agree with this evaluation, and I deplore it as much as you do. It is true that the textbooks tend toward a certain aridity and are pedantic. I have nothing else to say.

A UNIVERSITY STUDENT. I am a student, but those problems, we see them, but we don't have time to deal with them. If we did, then we would still be at the beginning when we get to be forty years old.

SERGE LANG. But nobody asks you to do that the whole year long. When you go to a concert, nobody asks you to do music all the time till you are forty.

THE STUDENT. During math class, we see interesting problems, but if we go deeper into them, we spend hours and hours, and there is a lot of other things to do. The curriculum is much too heavy to allow us to take an interest in things like that.

SERGE LANG. It depends on the level. I think the curriculum is filled with stuff that could easily be taken out without anybody missing it. [*Laughter.*]

STUDENT. Can you tell me which ones?

SERGE LANG. Bring me the book and I'll show you. You can find more and more technical exercises, which don't teach anybody anything.[10]

---

[10] Here I misunderstood. I am speaking of elementary and high schools. Beyond, that is at the student's level, the situation is different, and complicated in different ways. I sympathize with what he said, but this is not the time to go into the contradictory requirements of education at the college level.

[*The preceding dialogue is extracted from a long general discussion—too general—on teaching. I pass now to my last answer.*]

I spend my life doing mathematics. From time to time, I do mathematics with you, just like this. I prefer to do this than to have general discussions. I prefer to come here, give this talk, show you how I teach, point my finger at you, and make you ask questions . . . and if it works, that was one of the ways of doing things. Maybe in this way, you will find your own inspiration, to do as you want to touch others. That's how I function, rather than by pontificating with generalizations. I don't like generalities. This does not mean that I never generalize, sometimes I do, but I don't like them.

There is some success in what I did today, for instance [*showing the high school student*] what's you name?

STUDENT. Gilles.

SERGE LANG. Gilles is one of those who asked questions on the mathematics. Others took refuge in pedagogical questions. I prefer Gilles' questions.

ANOTHER HIGH SCHOOL STUDENT. [*Antoine, who had also come last year.*] You told us that the formulas

$$x = \frac{1 - t^2}{1 + t^2} \quad \text{and} \quad y = \frac{2t}{1 + t^2}$$

give all the rational solutions of $x^2 + y^2 = 1$, except $x = -1$ and $y = 0$. Can you give us the proof now?

SERGE LANG. Yes, naturally, I had even hoped somebody would ask that question earlier. The proof is easy. Suppose that (x, y) is a rational solution. Let

$$t = \frac{y}{x + 1} \, ;$$

and don't ask me where it came from, with a little ingenuity you could discover it yourself.[11] We then have

$$t(x + 1) = y,$$

and squaring, we find

$$t^2(x + 1)^2 = y^2 = 1 - x^2 = (1 + x)(1 - x).$$

---

[11] G. Lachaud informs me that Diophantus, and therefore the Greeks, had not raised the question whether the formulas give all the solutions. He also informs me that this result is due to the Arabs of the 10th or 11th century [La–Ra]. The algebra necessary to prove this result is approximately at the same level as the algebra used by Diophantus, and so we see a posteriori that once the question is raised, one finds the answer rather easily.

You can then cancel $x + 1$ on both sides, and we find

$$t^2(x + 1) = 1 - x.$$

Therefore $t^2x + t^2 = 1 - x$, and

$$x(1 + t^2) = 1 - t^2.$$

Divide both sides by $1 + t^2$ to find the formula

$$x = \frac{1 - t^2}{1 + t^2}.$$

One more line will give you the corresponding formula for y.

You can interpret the argument geometrically, thanks to ideas which appeared only in the 17th century, namely coordinates and the representation of equations by curves. Namely, $y = t(x + 1)$ is the equation of a straight line, passing through the point $x = -1$, $y = 0$; and whose slope is equal to t. This line intersects the circle of radius 1 at the point (x, y) such that

$$x = \frac{1 - t^2}{1 + t^2} \quad \text{and} \quad y = \frac{2t}{1 + t^2},$$

which is precisely what we have just shown.

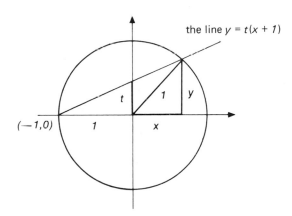

I would like to add a few words on the difference between integral and rational solutions. We have seen that an equation like

$$y^2 = x^3 + ax + b$$

can have an infinite number of rational solutions, obtained as multiples
nP of some rational point P. For instance, in the example

$$y^2 = x^3 - 2,$$

we started with the integral point $P = (3, 5)$. One can prove that it is the
only integral point on the curve. Furthermore, there exists a very general
theorem of Siegel, which says that the number of integral points on a
curve $y^2 = x^3 + ax + b$ is always finite [Sie].

When a, b are integers, then any point of finite order (x, y) must necessarily be an integral point, that is x, y are integers, according to a theorem
of Lutz–Nagell. Of course, the converse is false, as in the example with
$x = 3$, $y = 5$ which is not of finite order.

By the way, about points of finite order, let me give you an easy exercise. Go back to the curve $y^2 = x^3 + 1$. We found the integral points:

$$x = 0, y = \pm 1; \qquad x = 2, y = \pm 3; \qquad x = -1, y = 0.$$

I told you that there were no other rational points. It follows that if you
take any one of these points, for instance $P = (2, 3)$, one of the multiples
nP with a suitable n, must give O. So I ask you to compute explicitly
$2P$, $3P$, $4P$, $5P$. It's easy with the addition formulas, and you can also do
it on the graph. You will find all the other integral points, and you will
also find that

$$5P = -P.$$

Therefore $6P = 5P + P = O$, and the point P has order 6.[12]

MR. BRETTE. [*Question asked two days later.*] You said that the order
of a rational point is at most equal to 12. But if you look at all real points,
does there exist points of arbitrary order?

SERGE LANG. Yes, and one can even describe them quite precisely.
Suppose first for simplicity that there is no oval. Then for each integer
$n \geq 2$, there exists one point P of order exactly n (that is, P does not have
smaller order), and such that every point of order n is equal to an integral
multiple of P. If there is an oval, then the situation is the same, up to a
point of order 2.

---

[12] I thank Mr. Brette for having drawn a very effective illustration of the curve (on the opposite page), which shows very clearly the point at infinity, and the rational points of finite
order. Note that P_1 has order 6, P_2 has order 3, and P_3 has order 2.

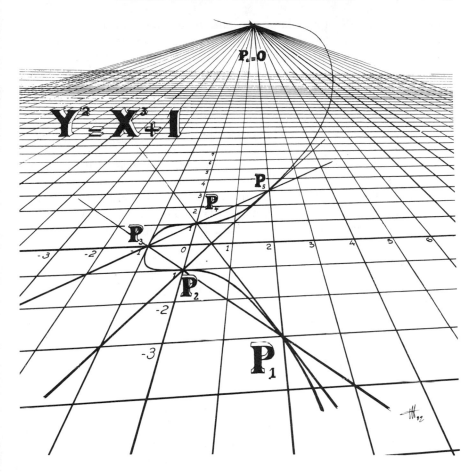

Added August 1982

Following the talk, I continued to think about the determination of
integral and rational points, to try to get more coherent conjectures about
them. Siegel's proof did not give an upper bound for the integral points,
depending on the coefficients a and b of the curve

$$y^2 = x^3 + ax + b.$$

We now suppose that a, b are integers. In the special case $y^2 = x^3 + b$,
Baker [Ba] has given effective bounds, although far from the best possible
ones which one might expect. For example, there is a conjecture of
Marshall Hall which says that when b is an integer, then x is bounded in

absolute value by b^2 times some constant, independent of b.[13] I think one might expect something similar in the general case. It would be very interesting to show that there is some constant k such that for any integral point (x, y), the integer x is bounded in absolute value by a constant times the maximum of the absolute values of a^3 and b^2 raised to the kth power. One can write this in the form

$$| x | \leqq C \max \, (\, | \, a \, |^{\, 3}, \, | \, b \, |^{\, 2})^k.$$

Finding bounds like that would constitute great progress in the study of such curves.

It would also be interesting to find bounds in the context of points of infinite order. More precisely, let $P = (x, y)$ be a rational point. Write $x = c/d$ as a fraction as we already have done. Define the height

$$h(P) = \log \max \, (\, | \, c \, |, \, | \, d \, | \,).$$

Considerations having to do with the Birch–Swinnerton–Dyer conjecture have led me to the following conjecture, understandable by someone who is not necessarily a number theorist. There exist points $P_1, \ldots, P_r$ as we have considered them previously, ordered by increasing height, such that

$$h(P_r) \leqq C^{r^2} \max \, (\, | \, a \, |^{\, 3}, \, | \, b \, |^{\, 2})^{1/12 \, + \, \epsilon}$$

where C is some constant, and ϵ approaches 0 as $\max \, (\, | \, a \, |^{\, 3}, \, | \, b \, |^{\, 2})$ increases indefinitely. See [La 2].

The existence of such bounds would allow an effective way of finding all the rational points, since these can be expressed by means of addition and subtractions starting with $P_1, \ldots, P_r$ and points of finite order.

Note that in tables, for instance that of Cassels or Selmer [Se], it seems that there is a better bound than that described above. If we let

$$H(P) = \text{maximum of } | \, c \, | \text{ and } | \, d \, |,$$

then one has an approximate inequality

$$H(P) \leqq \max \, (\, | \, a \, |^{\, 3}, \, | \, b \, |^{\, 2})^k$$

with $k = 1, 2,$ or 3. I give a numerical example taken from Selmer's table, where he considers the related equation

$$X^3 + Y^3 = DZ^3$$

---

[13] Recall that the absolute value of a number is the positive part of the number. For example, the absolute value of -3 is 3, and the absolute value of -3 is 3 also. The absolute value of x is denoted by $| \, x \, |$.

as in Fermat's equation. Mr. Brette used the computer to transform Selmer's biggest solutions back to the form we have considered, that is

$$y^3 = x^3 + 2^4 3^3 D^2$$

with $b = 2^4 3^3 D^2$.

Take $D = 382$. Then we have a solution $x = u/z$, where

$$u = 96,793,912,150,542,047,971,667,215,388,941,033$$
$$z = 195,583,944,227,823,667,629,245,665,478,169.$$

The reader can compare this solution with b^2. You will find that $u \leq b^6$, so $k = 3$ works. It would be interesting to make a statistical analysis of such polynomial bounds, rather than the logarithmic bounds conjectured previously.

Appendix

I reproduce below a table of Cassels [Ca]. The following comments will describe the content of the table, and how to read the columns.

Given a curve

$$y^2 = x^3 - D \qquad \text{with} \qquad -50 \leq D \leq 50,$$

we look for rational points $P_1, \ldots, P_r$ on the curve such that for every rational point P, there exist integers $n_1, \ldots, n_r$ uniquely determined by P, such that

$$P = n_1 P_1 + \cdots + n_r P_r + Q,$$

where Q is a point of finite order. Therefore, r is the rank.

In all cases, we have $r = 0, 1$, or 2. For instance, let

$$P_1 = (x, y).$$

Rather than make a table of rational numbers, we prefer integers. So we express the rational numbers x, y as fractions,

$$x = u/t^2 \qquad \text{and} \qquad y = v/t^3$$

with integers u, v, t. The equation of the curve can be expressed in terms of u, v, t in the form

$$v^2 = u^3 - Dt^6.$$

"None" means that the rank is equal to 0, and so the only rational points are of finite order, if there are any.

The first column gives P_1 if it exists.

The second column gives P_2 if it exists, besides P_1.

Table 1

$$v^2 = u^3 - Dt^6$$

D	P_1			P_2		
	u	v	t	u	v	t
1		None				
2	3	5	1			
3		None				
4	2	2	1			
5		None				
6		None				
7	2	1	1			
8		None				
9		None				
10		None				
11	3	4	1	15	58	1
12		None				
13	17	70	1			
14		None				
15	4	7	1			
16		None				
17		None				
18	3	3	1			
19	7	18	1			
20	6	14	1			
21	37	188	3			
22	71	119	5			
23	3	2	1			
24		None				
25	5	10	1			
26	3	1	1	35	207	1
27		None				
28	4	6	1			
29	3,133	175,364	3			
30	31	89	3			
31		None				
32		None				
33		None				
34		None				
35	11	36	1			
36		None				
37		None				
38	4,447	291,005	21			
39	4	5	1	10	31	1
40	14	52	1			
41		None				
42		None				
43	1,177	40,355	6			
44	5	9	1			
45	21	96	1			
46		None				
47	12	41	1	63	500	1
48	4	4	1			
49	65	524	1			
50	211	3059	3			

Table 1 (cont.)

$$v^2 = u^3 - Dt^6$$

D	P_1			P_2		
	u	v	t	u	v	t
−1		None				
−2	−1	1	1			
−3	1	2	1			
−4		None				
−5	−1	2	1			
−6		None				
−7		None				
−8	2	4	1			
−9	−2	1	1			
−10	−1	3	1			
−11	7	19	2			
−12	−2	2	1			
−13		None				
−14		None				
−15	1	4	1	109	1,138	1
−16		None				
−17	−1	4	1	−2	3	1
−18	7	19	1			
−19	5	12	1			
−20		None				
−21		None				
−22	3	7	1			
−23		None				
−24	−2	4	1	1	5	1
−25		None				
−26	−1	5	1			
−27		None				
−28	2	6	1			
−29		None				
−30	19	83	1			
−31	−3	2	1			
−32		None				
−33	−2	5	1			
−34		None				
−35	1	6	1			
−36	−3	3	1			
−37	−1	6	1	3	8	1
−38	11	37	1			
−39	217	3,107	2			
−40	6	16	1			
−41	2	7	1			
−42		None				
−43	−3	4	1	57	2,290	7
−44	−2	6	1			
−45		None				
−46	−7	51	2			
47	17	89	2			
−48	1	7	1			
−49		None				
−50	−1	7	1			

Bibliography

[Ba] A. BAKER, "Contributions to the theory of Diophantine equations II: The Diophantine equation $y^2 = x^3 + k$", *Phil. Trans. Roy. Soc. London* A **263** (1968), pp. 173–208.

[B–SD] B.J. BIRCH and P. SWINNERTON-DYER, "Notes on elliptic curves I." *J. Reine Angew. Math.* **212** (1963), pp. 7–25.

[Ca] J.W. CASSELS, "The rational solutions of the diophantine equation $y^2 = x^3 - D$," *Acta Math.* **82** (1950), pp. 243–273.

[Did] DIDEROT, article "Dimension", *Encyclopédie* Vol. 4 (1754), p. 1010.

[Di] DIOPHANTE D'ALEXANDRIE, *Les six livres arithmétiques et le livre des nombres polygones*, Paul ver Eecke, Albert Blanchard, Paris 1959.

[Go] D. GOLDFELD, "Conjectures on elliptic curves over quadratic fields," à paraître.

[Ha] M. HALL, "The diophantine equation $x^3 - y^2 = k$," *Computers and Number Theory*, Academic Press, 1971, pp. 173–198.

[La–Ra] G. LACHAUD and R. RASHED, Une lecture de la version arabe des "Arithmétiques" de Diophante; cf. les *Oeuvres de Diophante*, Collection Guillaume Budé, Les Belles Lettres, Paris, 1984.

[La 1] S. LANG, *Elliptic Curves: Diophantine analysis*, Springer-Verlag, 1978.

[La 2] S. LANG, "Conjectured diophantine estimates on elliptic curves," in a volume dedicated to Shafarevich, Birkäuser, Boston–Basel, 1983.

[Ma] B. MAZUR, "Modular curves and the Eisenstein ideal," *Pub. Math. IHES*, 1978.

[Mo] L.J. MORDELL, "On the rationnal solutions of the indeterminate equation of the third and fourth degrees," *Proc. Camb. Phil. Soc.* **21** (1922) pp. 179–192.

[Ne] A. NERON, "Quasi-fonctions et hauteurs sur les variétés abéliennes," *Ann. of Math.* **82**, No. 2 (1965), pp. 249–331.

[Poi] H. POINCARÉ, "Arithmétique des courbes algébriques," J. de Liouville, 5e série, t. VII, fasc. III (1901), pp. 161–233, Oeuvres complétes, t. V, Gauthier-Villars, 1950.

[Pod] V.D. POSDIPANIN, "On the indeterminate equation $x^3 = y^2 + Az^6$," *Math Sbornik*, **XXIV** (66), No. 3 (1949), pp. 392–403.

[Si]   C.L. SIEGEL, "The integer solutions of the equation $y^2 = ax^n +$ $bx^{n-1} + \cdots + k$," *J. London Math. Soc.* **1** (1926), pp. 66–68 (under the pseudonym X).

[Tu]   J.B. TUNNELL, "A classical diophantine problem and modular forms of weight $3/2$." *Invent. Math.* (1983) pp. 323–334.

[vN]   J. von NEUMANN, "The role of mathematics in the sciences and in society," address to Princeton Graduate Alumni, *Complete works*, vol. VI, pp. 477–490.

Great problems of geometry and space

28 May 1983

Summary: *To do mathematics is to raise great mathematical problems, and try to solve them. Eventually to solve them. This time, we shall treat problems of geometry and space, and we shall classify geometric objects in dimensions 2 and 3. Dimension 2 is classical: it's the classification of surfaces, which are obtained by attaching handles on spheres. One can also describe surfaces by using the Poincaré-Lobatchevsky upper half plane. What happens in higher dimensions? In dimension $\geq$ 5, Smale obtained decisive results in 1960. Last year, Thurston published great results in dimension 3. He conjectured the way such objects can be constructed starting with simple models, and also how one could obtain them from the analogue of the upper half plane in 3 dimensions. He proved a good part of his conjectures. We shall describe Thurston's vision.*

First Part: Rubber geometry.
   Curves, surfaces, equivalences, octopusses, sums of geometric object.

Second Part: The geometry of distances.
   Euclidean geometry, non-euclidean, distances, motions, translations, rotations, symmetries, identifications.

The link between the two and Thurston's conjecture.

It is uncommon on a Saturday afternoon in May to see 230 persons come, not only to listen to a conference on mathematics, but also to participate, answer questions, in short think about mathematics and get pleasure out of it. To be sure, the enthusiasm of the lecturer, the energy which comes from him, and the care which he exercises to explain his subject and ideas can hardly leave an audience insensitive. On the other hand, it seems clear that the pleasure is shared. First of all by me, but also by Serge Lang. One sees in him what should be natural for any good teacher, satisfaction in the face of positive reactions by the public, and the relevance of the questions which come from his audience, especially by some high school students. After the success of the first two conferences, one can easily understand that I wanted to invite him again, and that he accepted, not without some hesitations because he said that it would be difficult to choose a genuine mathematical topic which would nevertheless be understandable by a broad audience. Two weeks later, he phoned me from Germany to tell me that he had found a possible geometric subject, but that he would have to learn it. To my question: "In which books?" he answered: "I don't know how to read . . . Or rather, I know how to read but I don't like it. In a book, not everything is of equal importance, but one doesn't know it until one has read everything. It goes much faster to ask a friend to explain this stuff. It's more lively, and I can ask questions."

During the course of the year, I then received successive versions of his talk, which testified to his concern for clarity and simplicity. But it is hardly necessary for me to say here that these versions were only pale sketches compared to the following text, which reproduces faithfully the tape recording of his marathon talk, which lasted over three hours.

J.B.

The conference

The first hour

This is the third time that I come here, to the Palais de la Découverte, to do mathematics with you. Mr. Brette invited me the first time, and it worked, so I came back.

I see quite a few people here who came last time, How many were here last year?

[*About fifty hands go up.*]

Good. I see Antoine over there, he's already been here twice, so he's quite faithful. Those who were here last year perhaps remember that just before starting the conference, I had looked at some high school book in Brette's office, and I had become quite upset because it was so lousy. It took me a good twenty minutes to get over it. I don't know if you noticed, today, before the conference, you heard a record of lute music, which is the music I like best. Brette put it on to calm me down. [*Laughter.*]

Two years ago, I did something on prime numbers, and last year, I did what's here on this reprint, what's called diophantine equations. And I asked people what mathematics meant to them. One lady told me: "It's to work with numbers." Well, those answers are for the birds, because this is not at all what it means to do mathematics. I wanted to show you what mathematics are about, what the great problems of mathematics are about, and why one gets excited about them.

Actually, in the first two talks, I did things which were related a little to algebra, and even a lot. In particular, last year, I wrote down some formulas, and then six persons immediately walked out, because formulas . . . well, people don't like them so much. But sometimes formulas are necessary. Still, I wondered whether it would be possible to do something without any formulas, without any connection to algebra, and without numbers. This means doing something geometric, in space, on problems having to do with geometric objects.

That's not the mathematics I usually do myself. Personally, I lean toward algebra and number theory. So I thought about it, as I left Paris for Bonn, in Germany, trying to figure out what I could talk about this year. I go to Bonn like this every year, for the last twenty, twenty-five years. Hirzebruch organizes a conference, and the people who go there are mostly interested in geometric subjects. I talked to some of them, and I realized that it would be possible to do one topic, on some recent research, discovered about a year ago.

It's very nice in Bonn. Mathematicians try to hold their conferences in pleasant surroundings, and there, we do mathematics between a glass of Rhine wine and a strawberry tart—strawberries are in season just at that time—and also a boat trip on the Rhine.

There is still some time left to do mathematics, and that's where I learned the subject I am going to talk about today, some recent

73

discoveries of a guy named Thurston. I learned it from Walter Neumann. We spent three hours in front of a blackboard, and he taught me what Thurston had done. And today, I pass it on to you.

I really don't know if I'll be able to go on like this, finding other subjects, because it isn't so easy. It must be real mathematics, done by real mathematicians. On the other hand, I must be able to explain it to a Saturday afternoon audience. And also, as in all aesthetic situations, somebody may like it, and somebody else may not like it. So it may not work out. It's a question of personal tastes, and the personal reactions which you may have toward any specific topic.

All right, so what I want to do, is to classify geometric objects. We are immediately motivated for this. After all, we live in a space with at least three dimensions, like this. But you all know that there may be more than three dimensions. So we want to describe the kind of space we live in, we want to know what it looks like. Locally, for instance in this room, it's a three-dimensional space.

As a model, it's OK for this room. But we already know that if you look very far out, it doesn't work. We know that the euclidean model is wrong. It works in restricted cases, but it does not apply in other situations. So what do physicists do? They try to find out which models are applicable. But a mathematician, that is, a pure mathematician, doesn't care whether the models he thinks about can be applied or not. He constructs nice models, geometric models, and if they are beautiful, that's what matters to him. He doesn't care whether these models can be used to describe the universe or not.

And we can do such geometric models in dimension 1, in dimension 2, in dimension 3, or also higher dimensions, like 4, 5, or whatever. I thought for a moment that I could do something today in higher dimensions, but I realized rapidly that I could not, at least in an hour and a half. It would have taken too much preparation. So I'll limit myself to dimensions 1, 2 and 3.

So, one-dimensional objects are like this, they are curves.

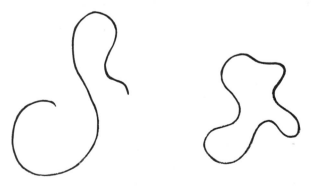

A LADY. What about a straight line?

SERGE LANG. A straight line is a special type of curve. Now if I take a circle, and other curves like that, they look like each other.

Should we consider them to be equivalent? What common properties do they have?

SOMEONE. They are closed.

SERGE LANG. Yes, they are closed, they turn around. If I take just an interval, like this:

then it doesn't turn around, it's an interval.

But the three curves are closed. For many properties, we don't want to distinguish between these three curves. So we say these curves are equivalent.

What does "equivalent" mean, in general? Well, I don't want to give a formal definition, but I can say informally that we suppose everything is made out of rubber. We are dealing with rubber geometry. We say that two objects are equivalent if by pulling in one direction, pushing in another, if these objects were made out of rubber, then I could deform one into the other. This gives me a notion of equivalence.

So if the curve is a rubber band, it's clear that I can deform it to the other curve, or that I can deform it into a circle. So these curves are equivalent. I use the sign $\sim$ to denote equivalence. So I can write:

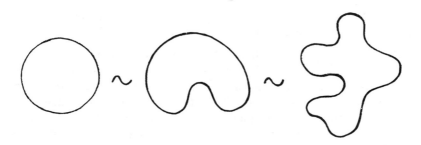

I can also draw something like this:

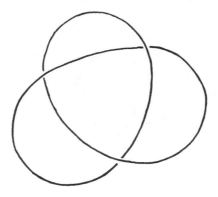

Is this thing closed or open?

THE AUDIENCE. It's closed.

SERGE LANG. Then is it equivalent to the others or not? Suppose it's a rubber band. Who says it's equivalent to the others?

[*Some hands go up.*]

Who says it is not equivalent?

[*Other hands go up.*]

Who keeps a prudent silence? [*Laughter.*] You, for instance. [*Serge Lang points to a lady in the third row.*]

THE LADY. It's not equivalent. There is a knot.

SERGE LANG. Yes, there is a knot. When I said that the other curves are equivalent, I could deform them into each other in the plane. I mean, if they were rubber bands, I could make the deformation entirely in the plane. But the knot, over there, it lives in three-dimensional space, and your intuition is right: I cannot deform it into the circle in three-dimensional space. In some sense, the knot is therefore different from the circle, and from the other curves. However, can you conceive a situation when I could deform the knot into a circle? Antoine, what do you say.

ANTOINE. [*The answer cannot be heard on the tape.*]

A LADY. Sometimes you can make two knots which are opposite to each other, and undo each other.

SERGE LANG. For now, the knot is in 3-dimensional space. But there is no reason to limit ourselves to this space. It is true that in four-dimensional space, we can deform the knot so that it becomes a circle. One can also prove that this is impossible in 3-space. Although we can rely on our intuition in 3-dimensional space, when proving things in higher dimensions, one should first write things down more rigorously, and our intuition becomes rather delicate. I also want to make you understand that things aren't that simple.

We now see that we can raise two different questions:

Can we deform the knot into the circle in 3-dimensional space?

Can we deform it abstractly, or in a higher dimensional space?

The answers are different, depending on the space in which we embed the knot.

At first, I didn't say where you could make the deformation, when I defined the notion of equivalence. Now I say that I allow deformations in spaces of arbitrarily high dimension, bigger than 2 or 3. So the dimension of the circle, which is 1, has to be regarded as entirely different from the dimension of the space in which we consider the circle.

Now I want to say something else about deformations. Take something which is not a circle, say an interval, like this, with or without its ends.

with the ends   without the ends

If I include the ends, then I say that the interval is closed. If I don't include the ends, then I say that the interval is open. Suppose the interval is made of rubber, and I deform it, like this. [*Serge Lang draws as he speaks.*]

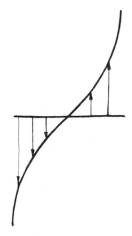

The points to the right, I move them up and the points to the left, I move them down. So I take the rubber band, and stretch it up as I go to the right, but faster and faster. And when I go left, I stretch it down, also faster and faster. Then we see that the interval is equivalent to a curve which goes arbitrarily far away, which extends to infinity as one sometimes says.

[*Someone raises their hand.*]

SERGE LANG. Yes?

A LADY. It's going to close up at infinity.

SERGE LANG. No, infinity is not a point. Take a line, like this:

This line does not close up.

A GENTLEMAN. If it's made of rubber, one can close it up. [*Laughter.*] It is not an interval, but with an interval, you can also make up a circle.

SERGE LANG. Watch out! If you close up the line, or if you close up the interval, then you have to put some points at the end over each other. Take an interval containing its end points. If I join up the end points, then I do get a circle.

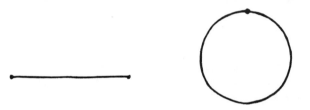

THE GENTLEMAN. But the interval can be deformed into a circle.

SERGE LANG. No, because if I deform it into a circle and I identify the two end points, then I do something which I don't want to allow in the definition of a deformation. I want to use the word in the sense that I do not allow identifications. If two points are distinct, then they must remain distinct during the deformation.

GENTLEMAN. But if you juxtapose them . . .

SERGE LANG. No, no. I don't want to! [*Laughter.*] It's a question of definitions. For the applications which I want to make here, I want to use the word "deformation" to mean that if two points are different, then they must remain different under the deformation. OK?

GENTLEMAN. Yes.

SERGE LANG. Good. Of course, there are other notions where identifications are allowed. In fact, in a short while, I shall discuss such notions and how to use them. But here, for deformations, I don't allow it.

I just wanted to show you this specific phenomenon, that I can deform an interval without its end points into an infinite band, which is itself equivalent to an infinite line. I can draw an equivalence between this infinite thing and the infinite line like this:

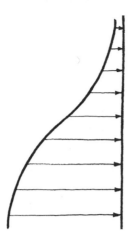

So I can straighten out the curve to a line. And that's the notion of equivalence that I want to work with.

All right, we've been talking about things of dimension 1. But already in dimension 1, we see that we can raise some problems. You might think that everything is known, but that is not the case.

Next, we look at dimension 2. It's going to get a little more involved. Objects of dimension 1 are called curves. Objects of dimension 2 are called surfaces. And there are surfaces with boundary and surfaces without boundary.

For an example of a surface, take the disc, the interior of the circle. If I put the circle together with its interior, then I get a surface with boundary. The circle is the boundary of the disc. So we can consider the disc as a surface without boundary if we leave out the circle, and with boundary if we include the circle.

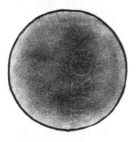

with boundary   without boundary

Now, if the disc is made of rubber, then I can represent it in other ways:

for instance, I can take the interior of a square. The boundary is then the perimeter of the square.

If everything is made out of rubber, are they equivalent?

THE AUDIENCE. Yes.

SERGE LANG. That's right, they are equivalent, I can take the disc and stretch it out to obtain a square, the interior of a square.

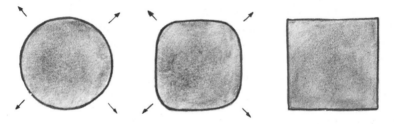

And the boundary of the disc, that is the circle, will become the boundary of the square.

[*A hand goes up.*]

SERGE LANG. Yes?

GENTLEMAN. But there is some difference, because of the derivatives.

SERGE LANG. Of course, there are corners. The gentleman says there is a difference, and he is quite right. There is a difference, but not from the point of view of rubber geometry. One can define other kinds of equivalence, for which the two objects would not be equivalent, because when I stretch out the disc and created a corner, then obviously this corner is not smooth. You could even say that from a certain point of view, the corner is disgusting. [*Laughter.*] It's not smooth, and it's not nicely curved. It's different in some sense. There is also a mathematical theory of corners, and now you see, we started from something rather simple, and already we can ask a lot of questions, which develop like a tree:

A person who walked out after about twenty minutes told the guardian: "I don't know if it's me who is not smart enough, but all this is just a farce."

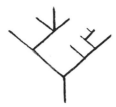

We climb up the tree, and we find two, or even several possibilities to go on. Depending on which equivalence relation you work with, you will find different answers for the same question. But right now, I only want to consider the rubber equivalence. Then the disc and the square are equivalent.

Of course, this does not depend on their size. I can make the square big or small. If it's made out of rubber, it will still be equivalent to the disc.

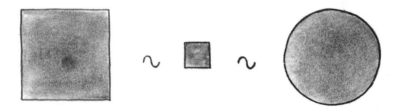

If I take just the interior of the disc, without the circle, then I get a surface without boundary, or the square without boundary. This is similar to the interval, without its end points. You remember the interval without the end points? Now I take the interior of the square, or of the disc which is equivalent to it, without the boundary, and the plane which extends to infinity in all directions. Do you think that the interior of the square is equivalent to the plane?

Who says yes?

A GENTLEMAN. The plane is indefinite?

SERGE LANG. It's the plane, yes, it's infinite.

GENTLEMAN. The square is without boundary?

SERGE LANG. Right, it does not have a boundary. I took it out, that's why I drew the dotted lines.

GENTLEMAN. Then it's also indefinite?

SERGE LANG. Like you say, it's indefinite.

GENTLEMAN. Then they are equivalent.

SERGE LANG. That's right. The square without boundary is equivalent to the plane. To summarize, every interval without boundary is equivalent to an infinite straight line, and every square or disc without boundary is equivalent to the whole plane.

But please note: if I take the square without boundary, I can still add the boundary if I want. Suppose however that I take a sphere, like this, the surface of a sphere:

It's a surface, but it does not have a boundary, OK? And if I stretch it, is it possible to stretch it in such a way that parts of it go as far away as you want?

GENTLEMAN. You can blow up the balloon indefinitely.

SERGE LANG. Watch out! I don't want to tear up the balloon. [*Laughter.*] The objects have to remain equivalent. I blow up the balloon, and punch it in or out some, like rubber, but I am not allowed to tear it up.

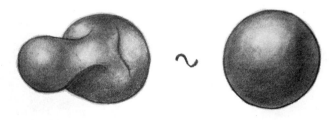

But if I stretch it, would it be possible to do what I did with the interval without its end points, to send parts of it as far away as you want? Who says yes?

[*Some hands go up.*]

Who says no? Actually, the answer is no. For instance, if I take the interval with its end points, would it be possible to stretch it so that it becomes equivalent to something infinite?

A LADY. It would be bounded.

SERGE LANG. That's right, one can prove that it is impossible. The interval with its end points is not equivalent to an infinite object.

LADY. Do you mean that the end points are fixed?

SERGE LANG. Oh no! They are not necessarily fixed, you can move them. For instance, it's equivalent to this thing here.

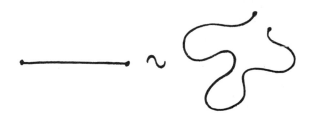

It suffices to pull, push, and stretch a little. But the problem is to find out if I can stretch faster and faster, as I did for the interval before. What happens if I stretch faster and faster, is that the end points have nowhere to go. Before, the points of the interval which came closer and closer to the extremities went higher and higher; or lower and lower. So to include the extremities in the deformation, I would need to tear off the end points. And I don't allow that.

A GENTLEMAN. You can put the end points at infinity.

SERGE LANG. No, we have to remain in the plane, there is no point at infinity in the plane, there are just points which go out as far as you want, it's not the same thing.

GENTLEMAN. Why is it forbidden?

SERGE LANG. It's forbidden in order to define the notion of equivalence. It's not forbidden in principle, it's not absolutely forbidden. You can add a point at infinity to the plane for other applications, but not for those I want to make today.

So you have to distinguish between things which have the property that under some deformations, some parts of them can be sent arbitrarily far away, and things which do not have this property. So let me write down a definition.

I say that something is compact if it contains its boundary (whenever the boundary exists), and if no deformation of this thing extends arbitrarily far away. In other words, if every deformation of this thing is bounded.

All of this is to come to the point of saying that the sphere is compact. Of course, the three-dimensional space in which we live goes to infinity . . . [*hesitating*], at any rate the naive model that we have in mind goes to infinity. But suppose you live on a sphere, and that you are very, very small. When you look around you, in any direction, it looks like a plane . . .

[*A hand goes up.*]

SERGE LANG. Yes?

A COLLEGE STUDENT. But the sphere is without boundary. You said: "compact without boundary".

SERGE LANG. Ah! if the surface has no boundary, it means that it contains its boundary. The terminology must accept this way of expressing yourself. If something doesn't have a boundary, then it can't help but contain its boundary, because there isn't any. [*Laughter.*] You must allow this possibility, because otherwise, you'll have a very hard time making simple mathematical statements.

Let's go back to people living on a sphere. Maybe they will see only a plane, even with good telescopes, and they will quickly come to the conclusion that the space on which they live is a plane. But suppose that a thousand years later, they make better telescopes, then maybe they will discover some curvature, they will see that space is curved, and they can start asking questions.

This is precisely what happened until Columbus. People thought everything was flat, except clever people, but there weren't so many of those.

THE AUDIENCE. So what's new! [*Laughter.*]

SERGE LANG. OK, so we look like that, and we can ask what happens if we keep on going, whether we can come back where we started from, or whether we go to infinity. The sphere is an example of something which is compact. If you start from some point, and keep going straight ahead of you in a given direction, then you come back where you started from.

Can you give me examples of other surfaces like that, compact surfaces?

THE AUDIENCE. A cube.

SERGE LANG. Yes, the surface of a cube, but it is equivalent to a sphere. Give me an example which is not equivalent to a sphere.

A GENTLEMAN. A torus.

SERGE LANG. What?

A GENTLEMAN. A torus.

SOMEONE ELSE. You make a hole in the sphere.

SERGE LANG. Are you a mathematician?

THE GENTLEMAN. A little.

SERGE LANG. That's already too much! I would like mathematicians not to intervene, because otherwise, it's cheating. [*Laughter.*] Of course, mathematicians know the answer, but I am not giving this conference for them. [*Serge Lang throws the chalk at the gentleman. Laughter.*]

So, you dig a hole, and you find this object, which has a hole in the middle.

Then one can show that this surface is not equivalent to the sphere, because of the hole. Now can you give me an example of a surface which is not equivalent either to the sphere or to the torus?

SOMEONE. A Klein bottle.

SERGE LANG. Some of you know too much.[1]

A CHILD. A pyramid?

---

[1] I don't want to go into this kind of technicality at this point.

SERGE LANG. No, that's equivalent to the sphere.[2]

A LADY. A box without its top?

SERGE LANG. Yes, but it will have a boundary. I want a surface without boundary. The preceding one doesn't have one, the sphere neither. I want a compact surface.

A COLLEGE STUDENT. You can make two holes, like eyeglasses.

SERGE LANG. There you are, that's what I wanted you to say. But are you a mathematician?

THE STUDENT. Yes.

SERGE LANG. Oh no, no, don't do this! Naturally, if you are a mathematician, you'll say: make two holes. But you are not playing the game. [*Laughter.*] That's why I am asking you not to intervene. I want to make people think for themselves.

So, you are right, I can make two holes. Like this.

And if I want still another example, what do I do?

A STUDENT. A torus with a knot.

A LADY. You can continue to make more and more holes.

SERGE LANG. Very good. You were here the first time madam? Two years ago? You don't remember? I remember you very well. Anyhow, you can make more and more holes. And there is no limit to the number of holes, except that there can only be a finite number.

---

[2] The audience has all kinds of people, including twelve year old children, high school and college students, engineers, and retired people. I learned subsequently that this child is 12 years old. Her teacher asked some students in her class who attended the conference to write up their impressions afterwards. This one wrote:

Of course, sometimes, I was a little confused, as when Mr. Lang asked for an example different from the sphere in dimension 2. I answered: "A pyramid", because I understood that Mr. Lang asked for a similar example. Otherwise, everything went well.

Another one said:

If I knew that what I say would be written down, I would have raised my hand more often.

So that's a theorem:

Compact surfaces, without boundary, are completely characterized, up to equivalence, by the number of holes. And there are no others.

I also have to make an additional hypothesis in the statement of the theorem. I should have said: an orientable surface. But I don't want to get into this kind of question now. So forget I said it. If I had not said it, then someone would have raised a fuss, some mathematician. [*Laughter.*]

[*As Serge Lang writes the statement of the theorem on the blackboard, space runs out, which forces him to erase another part of the blackboard.*]

So, there is no space left! Then you must all write to the Secretary of Education, so that he gives more funds to Mr. Brette for the Palais de la Découverte, so he can get more blackboards, and bigger ones, in a big room, and so on . . . Non-compact funds, if possible. [*Laughter.*] You all write to the Secretary, after the conference. I write the theorem:

Surfaces which are compact, without boundary and orientable (just for my conscience) are characterized up to equivalence by the number of holes.

That's the general model for surfaces.

Now let's look at surfaces with boundary.

A GENTLEMAN. And when there are only holes left?

SERGE LANG. There always remains some surface. I am doing all this as an introduction to objects in three dimensions, when it's going to become much more serious.

OK, now I draw a surface with boundary. Someone had already mentioned a cylinder.

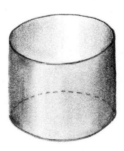

What's the boundary of the cylinder?

A GENTLEMAN. A circle.

SERGE LANG. Right, there is a circle on top and a circle on the bottom. The boundary of the cylinder is composed of two circles.

A LADY. There are also some edges.

SERGE LANG. No, because when the cylinder turns around, and you look at it sideways, you won't see these edges.

LADY. Then there are two boundaries?

SERGE LANG. Yes, or rather, there is a single boundary composed of two circles. Nobody said that the boundary has to consist of only one piece. It doesn't have to be connected.

Now I'll draw one which is a little more fun. Who can tell me how to draw a surface with a boundary consisting of more than two pieces?

GENTLEMAN. A face.

SERGE LANG. Yes, for instance.

LADY. A sieve. [*Laughter.*]

SERGE LANG. Yes, very good. Let me draw another one.

What is this?

GENTLEMAN. An upside down vase.

OTHERS IN THE AUDIENCE. A pair of pants.

SERGE LANG. Yes, a pair of pants. The boundary consists of the circle on top, and the two circles on the bottom. So the boundary has three pieces.

Now I am going to do something that mathematicians like a lot. Mathematicians like to combine things to make sums. Suppose you have two pairs of pants.

What can I do to them? If I take a circle on each one of them, I can glue them together.

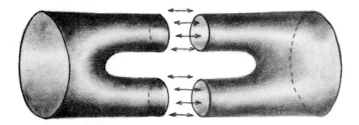

And I can do the same thing with the other leg. Then I obtain something which I can call a sum of the two pants.

GENTLEMAN. But you don't have the right to do this, you are identifying things.

SERGE LANG. Now I have the right. I am doing sums, I am sewing the pants together. [*Laughter.*] I have the right to sew. We are coming to the point when I have the right to identify.

LADY. Why didn't you have the right to identify a while ago, but now you have the right to identify?

SERGE LANG. You always have the right to identify, to put two points together. Do what you want. But for what purpose? To define the notion of equivalence, you don't have the right. I did not say that when I identify, then I obtain an equivalence. I said I obtain a sum. It's not the same thing. For equivalences, I am not allowed to identify points. For sums, I do have the right. To do a sum consists in identifying pieces of the boundary.

So if I draw the sum, I obtain something like this, with a hole, and still a boundary which contains two circles.

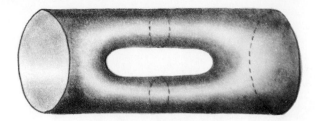

Now I can again do a sum, to eliminate the circles.

GENTLEMAN. In any case, you identify the boundaries of two distinct things.

SERGE LANG. Yes, two distinct objects. I take two pairs of pants, yours and mine, and I sew them together. [*Laughter.*]

Now I could also make a sum by taking a single pair of pants, and by identifying the two pieces at the boundary of the legs. It also gives me a surface with a hole.

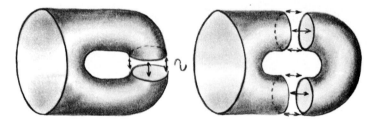

It's the same thing as if I had taken the sum of the surface with a cylinder, it's equivalent. I would get a hole, but there remains a single circle as the boundary.

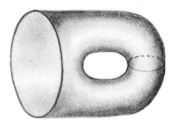

Now this circle, I want to eliminate it. How do I do that? What sum do I have to take to eliminate the boundary completely?

GENTLEMAN. A half-sphere?

LADY. A cover.

SERGE LANG. Precisely. A cover, which is a disc with its boundary.

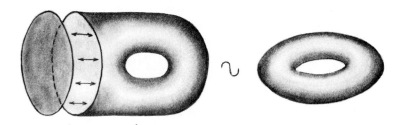

You glue them together, and you get something without boundary. That's what I wanted to show you. If I take surfaces with boundaries, with circles as their boundaries, and I take their sum, I can add them a certain number of times, and get a surface with holes, but without boundary.

Take the pants again. I can eliminate the boundary first by sewing the bottoms together, and then putting a cap at each end. I get a torus.

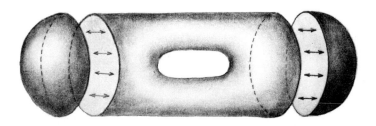

A STUDENT. But you could also have sewn them together along the belts.

SERGE LANG. Yes, I could, but that would have created a new hole.

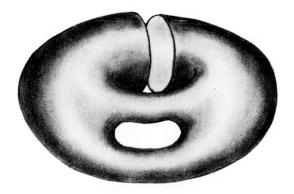

Mathematicians love to do that, it's one of the things they get a high on. [*Laughter.*] If you get a high by making pants and sewing them together, then by definition this is called doing topology, and you are a topologist.

Even if you don't have a boundary, you can create one to define still another kind of sum. Up to now, we did only sewing, but you can also do some surgery. Take a surface, like this, nice and smooth, without boundary. Then I cut off a disc.

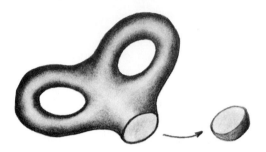

This yields a boundary, a circle which was not there before. I want to do the same thing with another surface, to create another circle. No I am in the previous situation, I have two surfaces with boundaries, and I can take their sum along these circles.

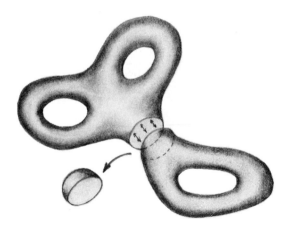

In this way I can define the sum of surfaces without boundaries. If I do this sum, between any surface and the sphere, then I find the same surface, up to equivalence. One can say that the sphere is the neutral element for this kind of sum.

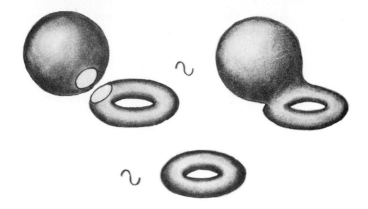

On the other hand, if I have a surface with two holes, and a surface with one hole (so a torus), and I take their sum, then I obtain a surface with three holes. If I take the sum of a surface with three holes and a surface with one hole, then I get a surface with four holes, and so on.

One says that a surface is irreducible if, when you express it as a sum of two surfaces, one of them is necessarily a sphere. The torus is an irreducible surface, and every surface which is not equivalent to the sphere can be expressed as a sum of toruses, a certain number of times, corresponding to the number of holes in the surface.

I repeat that in everything I have said, I meant the surfaces to be orientable. And we have just done part of the theory of such surfaces, which have dimension two.

Now I want to go on to objects having dimension three.

A while back, we spoke of people living in two dimensions, say on a surface. They are very small. What they see around them is also small, and it looks like a plane. But they could ask themselves: if we were able to see very far out, what would space look like? What about us? We are very small beings, on something which is three-dimensional. Are we living on something which is the analogue of the three-dimensional sphere? What happens if we look far out in space, do we find a hole? One can also ask the question in dimension two, but for us dimension three is more relevant.

We see a three-dimensional space, and we have telescopes which are more and more powerful. If we can see sufficiently far, what are we going to find? Are we living on an object equivalent to a sphere? Or are we going to find holes? This is getting serious. You can really raise this question about the nature of the universe. So if you are dead set on wanting a physical interpretation for what I am doing today, there you are.

I started in dimension 2 because it was easier to define the notion of sum than in dimension 3.

A GENTLEMAN. But the pants had dimension 3.

SERGE LANG. No, no! The surface of the pants has dimension 2. Of course, the pants exist in a three-dimensional space, but the surface itself has only two dimensions. You must distinguish between the dimension of the object itself, the surface itself, and the space in which it is embedded. Now it's the objects themselves which have dimension three.

Take the ball, for instance, the interior of the sphere, the full ball. It has dimension three.

the sphere   the ball

This ball, without the sphere which is its boundary, is equivalent to the whole three-dimensional space, for the same reason that the interior of the disc is equivalent to the plane R^2, or the interval without boundary is equivalent to the line R. The letter R denotes the real line, the straight line, and I put a small 2 on top to indicate that the space has dimension 2. For three-dimensional space, I would write R^3.

Something of dimension 1 is a curve. Something of dimension 2 is a surface. What do you call something of dimension 3?

AUDIENCE. A volume.

SERGE LANG. If you wish. But the word "volume" has several meanings. It can mean space itself, or it can mean the numerical value of this space. For instance, the interior of a suitcase, you might say that it is a volume, but you could also say that it is three cubic feet. You have to distinguish the two notions.

In rubber geometry, I don't measure a volume with numbers, because something can be equivalent to something else which is much larger, just by pulling and stretching.

I could go on talking about three-dimensional things, but they have a name in mathematics, a technical name. They are called manifolds, three-dimensional manifolds. I don't like this name, but that's the way they are called. Again I have the notion of rubber equivalence; I have the notion of compact manifold in other words, a manifold which does not go to infinity no matter how you deform it. I also have the notion of boundary, which will be what?

LADY. A surface.

SERGE LANG. Right, perfect. You have understood what I'm talking about.

OK, I've been talking for an hour. For the last two years, we stopped after an hour, then there were some questions, and people stayed around for quite a while. But I have to allow people to go if they want to. So we can have an intermission for a few minutes. The main topic which I want to discuss has to do with the classification of three-dimensional manifolds, and even of some non-compact objects.

In dimension two, I gave a theorem classifying surfaces: There is only the sphere and toruses with more and more holes. In dimension three, it's an extremely difficult problem, which mathematicians try to solve. This is precisely Thurston's contribution, to have stated a conjecture which describes all of them. There will also be sums, and there will be holes, but it will be much more complicated. That's what I want to do later.

But for the moment, intermission or recess, depending on your rhetoric.

[*Applause. Someone asks if he has the time to go get a drink across the street, and I answer yes. We start again after about fifteen minutes.*]

The second hour

[*At the start, the room was full, with about 230 persons. About three fourths have now come back for this second part.*]

[*On the blackboard, one can see the following picture, drawn by someone in the audience.*]

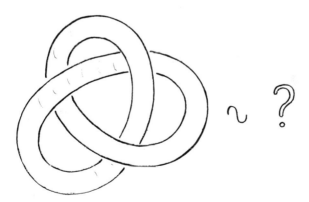

SERGE LANG. [*Looking at the picture.*] Ah, very good drawing. It's analogous to the knot, but with a torus. Do you have any questions on what I have done so far?

GENTLEMAN. Is this surface equivalent to a torus?

SERGE LANG. Very good question. What do you think?

SOMEONE. How many holes does this surface have?

SERGE LANG. Well, it's a surface with one hole, embedded in three-dimensional space, and one can't deform it on the torus if you ask that the deformation take place in three-dimensional space. But you can deform it on the torus if you allow higher dimensional spaces. It's just like the knot at the beginning. You see, there are several ways of representing the same surface.

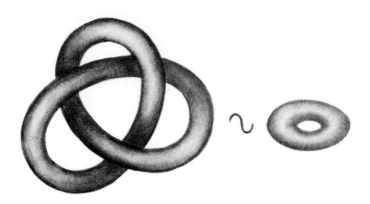

GENTLEMAN. If someone walks inside the surface, then they have no way of knowing if the thing is knotted or not.

SERGE LANG. Yes, excellent remark. It's just like the knot. If you move forward on the knot, always walking ahead, then you come back to where you started from, but you don't have any way of knowing that you are not on the circle.

[*A hand goes up.*]

SERGE LANG. Yes?

A LADY. What about the Moebius strip?

SERGE LANG. I have already said that I want to consider only orientable objects, precisely to eliminate this kind of thing, because I wanted to avoid technicalities to make simpler statements. It was to protect myself against someone who would complain that I was being incomplete. If I discuss non-orientable surfaces, then I won't have time to talk about three-dimensional things, and that's what I want to talk about. OK, the Moebius strip, many of you probably have heard about it, and there isn't much point dealing with it now. But you probably don't know so much about three-dimensional objects.

Besides, they are relevant for the world in which we live. I have already said that mathematicians work with lots of possibilities, lots of models. As mathematicians, we are interested in the beauty of these models, and not necessarily in their physical applications. Today, I classified surfaces, and I am interested in the classification of three-dimensional manifolds. I am trying to describe them all. After we know them all, then we can ask which ones correspond to the physical world, the world we live in. A physicist chooses among these models to find those which fit the empirical world. I myself have never done physics, and it disturbs me that there is a correlation between the world of experience, the world with which we come into contact with our senses. I have always felt that way, ever since I was a student. I have no ability in physics, which does not really interest me.

A LADY. No wonder students have a hard time to apply what they know in mathematics in order to do physics.

SERGE LANG. I have no reason to hide my personal tastes, but I don't impose them as being a law of nature. I like the classification of things, just like that, I make up models, and I tell the physicist: "Pick the one that suits you." On the other hand, there are other mathematicians, like Atiyah, or Singer, who are directly interested in physics. Conversely, there are physicists who understand mathematics very well, and who do both at the same time. And I am all for it. I make that quite clear to students, and I encourage them to do both if they are able to do so, and if they like it. But everyone has his own limitation.

OK, let's return to three dimensions. It becomes a lot harder to draw, because for instance, even the three-dimensional sphere, I can't show it to you. The ordinary sphere, the two-dimensional one, I could draw it as the set of all points which lie at a certain distance from a given point, which is called its center. The three-dimensional sphere S^3 again can be defined as the set of points in four-dimensional space, which are at a given distance from the center. So the sphere S^3 is embedded in four-dimensional space, and we can't draw it. But we can conceive it.

What I can do, however, are drawings which suggest what happens. Or give other representations. For example, in the plane, take two axes and points P, Q on these axes.

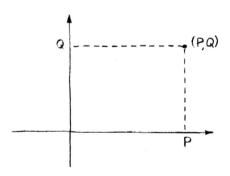

I can write the point (P, Q) where Q is on a line R and (P, Q) is in the plane. This construction is called a product. It is as if I put a line above each point P, and the point Q wanders along a line above P.

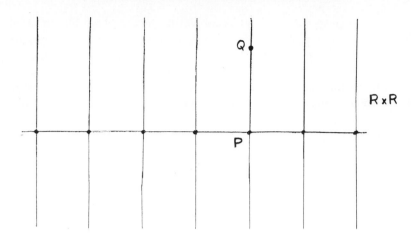

As I said, this construction is called taking a product. We saw that the plane R^2 is a product of R with R, and we write

$$R^2 = R \times R.$$

Similarly, if I have two intervals I_1 and I_2, and if I take the set of all points P in the first and Q in the second, then all the pairs of points (P, Q) like this constitute all the points of a rectangle.

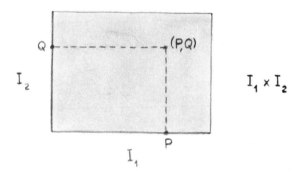

I can construct products like this with any two sets. If I take a surface F having dimension two, I can take its product with any one-dimensional thing.

If F has dimension 2, then the product is a manifold of dimension 3. It is the set of points (P, Q) where the first point P is a point on the surface F, and the second point Q belongs to a one-dimensional space.

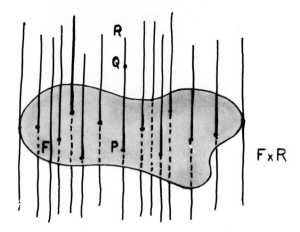

Let me draw another example. Let S^1 be the circle. I take the product $S^1 \times S^1$, that is the set of all couples (P, Q) where P is on the circle, and Q is on another circle. To each point P of the first circle, I can associate all the points on the other circle.

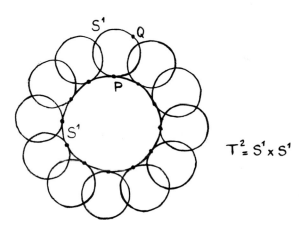

What kind of a surface do I get?

AUDIENCE. A torus.

SERGE LANG. Right, a torus. $S^1 \times S^1$ is a torus. I'll call it T^2, with T like torus, and the 2 because it has dimension two.

This notion of product allows me to construct higher dimensional object, and I can write them down. I don't need to draw them any more.

Now I can get to bigger dimensions. Take T^2 for instance, and its products with R, its product with a straight line. I would have a hard time drawing it, but I can represent it by drawing a torus, straight lines like that, and I can consider the torus as a section of this thing, this three-dimensional product.

$T^2 \times R$

Of course, this drawing is not correct, but it gives you a good idea of what's happening.

Next I want to draw more complicated things in three dimensions. I already have the sphere in three dimensions, and the product $T^2 \times R$, but I want things which correspond to surfaces, when I had holes. What does such a thing look like?

LADY. It could be a pipe with thick walls.

SERGE LANG. That's right. I want holes, toruses, and things which go to infinity. [*Serge Lang draws the following picture.*]

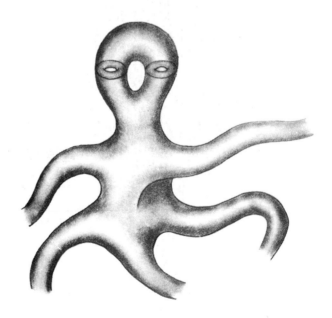

[*Laughter.*] So, here is a three-dimensional thing. What do I call it?

AUDIENCE. An octopus!

SERGE LANG. Precisely, an octopus. In dimension two, I had pants. In dimension three, I have octopusses. This suggests . . . suppose I take a pair of scissors, and I cut one of the legs. What do I get?

AUDIENCE. . . .

SERGE LANG. The octopus does not have a boundary.[3] If I cut one of its legs, I get something whose boundary will have dimension two, and which will be a torus.

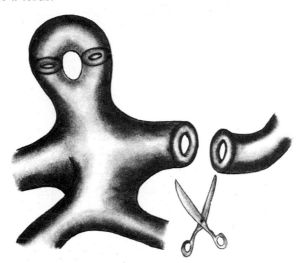

Now suppose you have some three-dimensional thing, whose boundary is a torus, and you have another thing whose boundary is also a torus. You must have an irresistible impulse to do something to them. What is your irresistible impulse? You [*pointing to a lady*].

LADY. To glue them together.

SERGE LANG. Right, just like before with circles. So I take two octopusses.

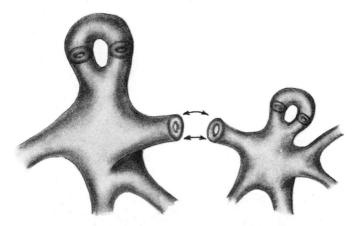

---

[3] Unfortunately, one cannot draw the picture correctly, and more than one could draw the sphere S^3. The way we have made the drawings, there is a boundary, but nevertheless, they show rather well what's going on.

I cut one leg of each one, I get two toruses, and I glue them together. Then I have taken the sum of the two octopusses along a torus.

I can also do a similar operation with only one octopus, by cutting two legs, and glueing the two sections, which are toruses.

Before we also had caps. What do we have now? I have a boundary which is a torus, and I want to eliminate the boundary. What do I do?

GENTLEMAN. You glue a ring.

SERGE LANG. Very good, that's right, a ring, the interior of a torus. I take the ring, and I glue it on the torus. It's the same type of operation as capping, but with one more dimension. So I have eliminated a piece of the boundary.

GENTLEMAN. How many legs can an octopus have?

SERGE LANG. Any number. Two octopusses can have a different number of legs, in which case they are not equivalent. [*Laughter.*]

GENTLEMAN. How do you take the sum of two octopusses if one of them has an odd number of legs, and the other has an even number of legs?

SERGE LANG. I didn't say that you had to glue all the legs of one to all the legs of the other. You can just glue together some of the legs, and then you can cap the rest of them with solid rings.

GENTLEMAN. What about $T^2 \times R$?

SERGE LANG. Well, $T^2 \times R$, it's ... euh ... it's like an octopus without holes, which has only two legs.

If I cut $T^2 \times R$ by making a section, then I get a boundary which is a torus. And there also exist octopusses without holes, with several legs.

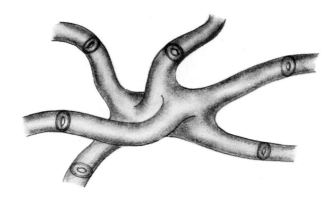

Just like for surfaces, one says that an octopus is irreducible if the only way to express it as a sum of two octopusses is when one of the two is equivalent to $T^2 \times R$, or when it is a capping operation, so glueing a ring like the gentleman said a minute ago.

If I take an octopus and take its sum with $T^2 \times R$, then I get an octopus which is equivalent to the one I started with. From the point of view of equivalence, I have not changed anything. One can say that $T^2 \times R$ is the neutral element with respect to this sort of addition, obtained by cutting and glueing a leg.

Now it is perfectly conceivable that after a finite number of additions like this, I can eliminate all the legs. Let me write this down.

After a finite number of additions, one can eliminate, in many ways, all the legs. Then one obtains a three dimensional manifold, compact, without boundary.

LADY. But there are holes.

SERGE LANG. Yes, definitely. We have eliminated the legs, but we have created holes, and there can be many of them. This is one of the ways of constructing three-dimensional manifolds, compact and without boundary . . . and orientable, just so my conscience does not bother me, and so nobody complains.

Of course, the next time you are on the beach, you can try it out [*Laughter*], take the legs of an octopus, and you can even knot them before you glue them together.

In order to classify octopusses, we must therefore classify the irreducible ones, and then we must classify the way you can add them together, as I did just now by cutting and glueing their legs.

Up to now, I have described geometric models: first models of surfaces, then models of three-dimensional manifolds, octopusses, the three-dimensional sphere S^3, which is not an octopus, and which is something else.

AUDIENCE. It does not have any holes.

SERGE LANG. Right, no holes. Then one can raise the following question.

Take all compact, three-dimensional manifolds, without holes and without boundary. Can you describe all of them? The problem is unsolved. Poincaré's conjecture is that a three-dimensional manifold, compact, without holes, without boundary is equivalent to the sphere S^3. Presumably there is no other. Of course, one should make more precise what is meant by a "hole", but let's leave this technicality aside for today.

Many people have tried to find the answer to Poincaré's conjecture, but so far no one has succeeded. In 1960, Smale proved the analogous conjecture in dimension bigger or equal to 5. After that, there remained the problem in dimension three and four. But the smaller the dimension, the harder it becomes, because you have not enough room to move around. The case of dimension 4 was just solved by Freedman in 1981. Many mathematicians contributed to the solution. They developed the theory as far as they could by "pure thought", without too many technical complications, and then they got stuck.

And it was Freedman, after six to ten years of work, who got it. It was very difficult, and very technical, and very complicated. It is one of the great result of contemporary mathematics, it is a first rate result.

There remains the three-dimensional case.

Therefore I cannot state a complete classification for three-dimensional manifolds, because Poincaré's conjecture is not yet proved.

For the other three-dimensional manifolds, there is a conjecture due to Thurston, of which he has proved a good part himself: it is possible to make up a concrete list, not too big, of certain manifolds, such that:

Every three-dimensional manifold, without boundary, compact and orientable, is either in this list, or is a sum of octopusses.

So far, I have carried out the part I wanted to do concerning rubber geometry. To make Thurston's conjecture more precise, and to describe the list more precisely, I have to deal with entirely different ideas.

And it is rather interesting, it is even very interesting, that the manifolds in this list will be constructed by the same method which will also allow us to construct octopusses with legs. In other words, we shall construct simultaneously, by the same process, manifolds without legs and manifolds with legs. To do this, we must leave rubber geometry, and do an entirely different kind of geometry. Most of you probably have already heard of it, non-euclidean geometry. But we have to do it in dimension three.

Before I go any further, do you have any questions? How do you feel about all this?

A GENTLEMAN. Through any point of an octopus, is there only one torus or are there several toruses?

SERGE LANG. It depends. If I cut a leg to make a section, then I get a torus. But if I cut elsewhere, it depends. I have to cut in the right place to get a torus, I have to cut a leg. A mathematician would say it this way:

> An octopus is a three-dimensional manifold, non-compact, without boundary, with a finite number of ends, each of which is equivalent to $T^2 \times R$.

So if I cut near a point, and not a leg, if I cut elsewhere besides an end which is equivalent to $T^2 \times R$, then there is no reason why I should get a torus. In fact, if I cut near a point, I can cut a ball, just like when you take an ice cream ball, and it leaves a boundary which is an ordinary sphere. You can also think of an air bubble in a piece of swiss cheese. It's the same thing as for surfaces. In that case, I cut a disc, leaving a circle as boundary, and that's how I defined the sum of two surfaces by glueing together two circles.

By cutting off balls, I can add together three-dimensional manifolds. I cut off a ball in the first, I cut off a ball in the other, this leaves a boundary in each one; a sphere in each one. I glue the two spheres together, and I obtain the sum of the manifolds. One says that a manifold is irreducible if, when I express it as a sum like this, one of the two must be equivalent to a sphere S^3.[4] In 1962, Milnor proved that every compact, three-dimensional manifold without boundary can be expressed as a sum of irreducible manifolds, essentially in a unique way.[5] This result reduces the classification of three-dimensional manifolds to the classification of irreducible manifolds. Always up to equivalence, of course.

Are there any other questions? No? OK, then we go on, and we come to the geometry of distances, and non-euclidean geometry. But I have been talking for two and a half hours. What do I do with the non-euclidean stuff? Do you want to leave? Have you had enough? I'll do what you want.

A LADY. No, we stay, you have stimulated our curiosity. We go all the way.

SERGE LANG. Oh, I have stimulated your curiosity! Then the octopusses sank in. Good [*laughing*], do you want another five minute recess and we go back to work?

AUDIENCE. No, we are all set, let's go on.

---

[4] Note that we use the word "irreducible" here with respect to the sum taken along spheres, while we already used this word when dealing with the sum along toruses. There are indeed two types of sums, and the context should always make precise which one is meant.

[5] J. Milnor, "A unique factorization theorem for 3-manifolds," *Amer. J. Math.* **84** (1962).

GENTLEMAN. Now that we are coming along, keep it up. [*And the gentleman makes a gesture meaning "keep it up".*]

SERGE LANG. OK, then, let's go. But you have some stomach to do mathematics! If any one wants to go, or has an appointment, don't be afraid to leave. [*Laughter.*] It's not that I want to kick you out, but still . . .

[*Several persons leave, and others will continue to leave during this last hour.*]

The third hour

Now I leave rubber geometry to do the geometry of distances. On the real line, or in the plane, or in ordinary three-dimensional space, we have the notion of distance. We are then interested in a new type of equivalence, which preserves distances.

I shall call motion a transformation which preserves distances. We shall have motions in euclidean geometry, and also in non-euclidean geometry, but I want to start with examples in the euclidean case, just to give you the idea. Using these motions, we can then do certain identifications, which will allow us to recover octopusses, and the geometry of distances will thus meet the rubber geometry. So we are going to do something quite substantial.

Let's start with the straight line R, with 1, 2, 3,..., -1, -2, -3, ...

Suppose given a certain direction, and a certain distance which I denote by an arrow.

Then take a point P. I can move it in the direction of the arrow, exactly this distance. Then I get a point Q which I call the translation of P, and which I write $\tau(P)$.[6]

$$P \longrightarrow Q = \tau(P).$$

For concreteness, take the arrow to have length 1. Then the translate of 1 is 2; the translate of 2 is 3; and so on. Now I identify a point P with its translations. Let me draw a point and its translations.

---

[6] The letter τ is a greek letter, tau. I would use a T except for the fact that we have already used T for a torus, so we need another letter.

If I identify like this, then I obtain a circle, just in the way that you wanted to do it at the beginning. If I take an interval, with its end points, and I identify the end points, then I get a circle.

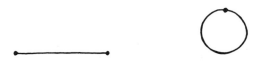

It was a very healthy reaction, a very mathematical reaction, for you to want to make these identifications, except that we are using them now. They were not allowed when we defined the notion of equivalence in rubber geometry. So, briefly, on the line, I get a circle by identifying a point with its translations in a given direction, by a given distance.

Of course, if P is a point, then I identify P with the next point $\tau(P)$, then with the next $R = \tau(Q)$. And how can I write R?

AUDIENCE. $\tau\big(\tau(P)\big)$.

SERGE LANG. Right, $\tau\big(\tau(P)\big)$, which I also write $\tau^2(P)$. And if I iterate a third time, then I write

$$\tau(R) = \tau\big(\tau(\tau(P))\big) = \tau^3(P).$$

And if I go in the opposite direction, then I write $\tau^{-1}(P)$.

All right, let's go to dimension 2. Then I have vertical translations, and horizontal translations, which I denote by τ_{ver} and τ_{hor}.

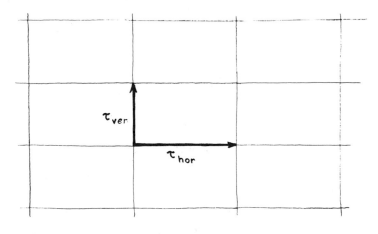

I can then make identifications, or translations, in two directions: the vertical direction, and the horizontal direction. Suppose that I identify the point P and the point Q in this next diagram. I identify the left hand side and the right hand side of the rectangle. And I also identify the top side and the bottom side.

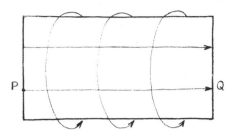

What do I get?

LADY. A sphere.

SERGE LANG. No! Watch out, to identify means what? When I identify the top and the bottom, then I get a cylinder.

Then if I identify the sides, what do I get?

AUDIENCE. A torus.

SERGE LANG. That's right, a torus, T^2.

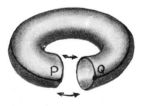

Now you see that I can describe the torus by means of a diagram in dimension two, by identifications and translations horizontally as well as vertically.

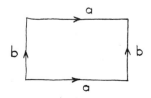

I can also make these identifications in the whole plane.

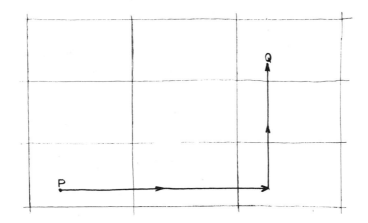

I'll say that two points of the plane arc equivalent if I can make horizontal and vertical translations which move one point on the other. But this is a different kind of equivalence from the rubber equivalence. Here I have the notion of direction and distance in addition.

I now need these two notions, which I had completely disregarded previously. So I have to specify which equivalence I mean, and I need two different words to denote these two equivalences. I have to fix some terminology, which I am going to do more systematically in a moment.

OK, I just got a torus, that is a surface with a hole, by making certain identifications. If I want a surface with several holes, can you guess what I should identify and how? Here I got a torus with a rectangle. If I want a surface with, say, two holes, what kind of identifications should I make?

LADY. Draw another line in the middle, or something like this.

SERGE LANG. Yes, you are right, one should draw more lines, but not quite as you said. Let me show you just what to do. Instead of four sides, use a polygon with eight sides.

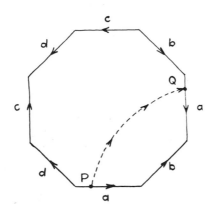

And make the identifications just as I drew them. For instance, I drew the point P identified with the point Q.

And if I want a surface with three holes?

AUDIENCE. Use a polygon with twelve sides.

SERGE LANG. Right, eight for the surface with two holes, and twelve for the surface with three holes. And if I want a surface with n holes, then I need . . .

AUDIENCE. $4n$.

SERGE LANG. That's right, and we draw it like this.

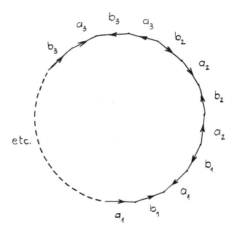

And that's how we get a representation of a surface with several holes.

GENTLEMAN. And if you have a polygon with six sides?

SERGE LANG. It won't give necessarily a surface with holes like the ones we had before. It can give something else, a non-orientable surface, but today, I want to limit myself to orientable surfaces.[7] But your question shows that you have understood what I am talking about.

You see, the torus T^2 can be obtained as a quotient of the plane, by means of certain identifications, which I am going to write with a slanted bar, on the left hand side:

$$T^2 \sim \text{Identif.} \setminus R^2.$$

These identifications were translations.

GENTLEMAN. What does the 2 mean?

---

[7] It all depends on the respective position of the sides which are to be identified, and of their orientation. In some cases one can find a torus, and in other cases, one can find a non-orientable surface. This is a good exercise: study those surfaces obtained by identifying sides in a polygon with $2n$ sides.

SERGE LANG. It's just to denote the dimension. The numbers which I write as superscripts, upstairs, are just to indicate dimension. I did not use numbers in any other way. I had sworn not to use numbers, but it is still useful to write the little 2 upstairs. I said I would do only geometric things, but the 2 denotes dimension. Do you allow this?

AUDIENCE. Yes.

SERGE LANG. Thanks. It's because I had promised not to use numbers. But this 2 is not really a number. [*Laughter.*]

Good, so I have represented T^2 as a quotient of R^2 by translations. And this was euclidean, with respect to translations. Now let's go to non-euclidean motions.

One of the models of the non-euclidean plane is the disc. I'll call it H^2, H for hyperbolic.

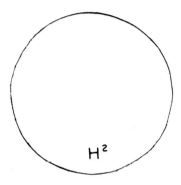

We need the notion of hyperbolic distance, and the notion of "line" with respect to this distance. In an audience like today's, there must be some of you who already know about this. Who knows already what this means?

AUDIENCE. ???

SERGE LANG. All right, I'll tell you what it means. By definition, a hyperbolic line is just an arc of circle in H^2, perpendicular to the boundary. I can draw it like this. Here are some hyperbolic lines.

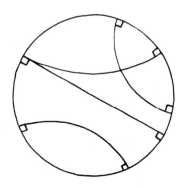

Some of them intersect each other, and others do not. Perpendicularity means the same thing as in the euclidean case.

As you see, you can have infinitely many lines passing through a given point P, but not intersecting another given line L. This cannot happen in the euclidean case.

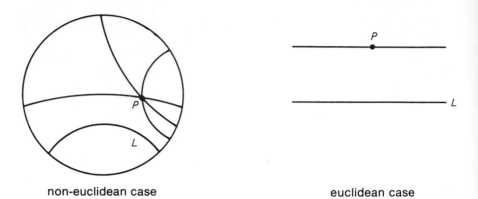

non-euclidean case   euclidean case

In euclidean geometry, given a line L and a point P, there is just one line passing through P and parallel to L.

We define a triangle just as in the euclidean case. Here is an example of a triangle, whose sides are line segments.

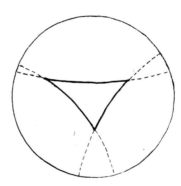

Now that we see what lines look like, we are going to describe the notion of hyperbolic distance, and the spaces we get from this point of view. After that, we'll make the connection with octopusses, and the classification of three-dimensional manifolds. I want to end up by stating Thurston's conjecture.

So we have to define a new distance, called hyperbolic distance. It's also called the Poincaré distance by the French, and the Lobatchevski distance by the Russians. I call it hyperbolic distance, so nobody gets upset. [*Laughter.*]

To describe this hyperbolic distance completely, I would need some formulas, and I don't want to write down formulas which are too technical. But I can speak of the rate of change of the distance when I start from the center, and move toward the boundary. This means that if r is the euclidean distance from the center,

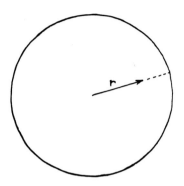

then the rate of change of the hyperbolic distance along a ray can be given by a very simple expression, which is

$$\frac{1}{1 - r^2}.$$

Here I supposed that the radius of the circle is equal to 1. So if I start from the center of the disc and I go toward the boundary along a ray, then how does the rate of change behave? You see, if r approaches 1, then r^2 also approaches 1, and $1 - r^2$ is very small. Then the fraction is very large. Therefore the rate of change of the distance becomes very large when I come near the boundary, and hence the distance becomes bigger and bigger. There is a formula which gives the distance in terms of r, with the logarithm. Who has heard of the logarithm?

[*Several hands go up.*]

Ah! Several of you do know. Then I'll give the formula. Those who don't know what the logarithm is don't need to listen. The hyperbolic distance along a ray is

$$d = \frac{1}{2} \log \frac{1 + r}{1 - r}.$$

So the distance becomes larger and larger as we get nearer to the boundary.

You see that this is analogous to Einstein's thing, and to the way the world is made up. Suppose we start at the center, and we start moving toward the boundary, as far as possible. What happens when we go very far out, in our own universe? We know that the euclidean model fails, we know that space is going to get curved, a little like the hyperbolic lines a while back. And we speed up. Suppose that a ray of light goes in the same direction. If I measure its speed, then I find 186,000 miles per second. Now suppose I go faster. If the model was euclidean, then when I measure again, I should find a smaller value for the speed of light. Right? Well, the answer is no, not at all! I always find the same value. If two trains were going at the same speed, in the same direction, then they would not move with respect to each other. But this does not work the same way with light. Light always travels at the same speed. And the reason is that as I go faster and faster, then I grow smaller and smaller, and my measuring apparatus will also grow smaller, and when I measure the speed of light, then I find a constant value.

In the hyperbolic plane, we meet an analogous phenomenon. Here I have drawn points at a distance of 1 unit between each other, along a ray.

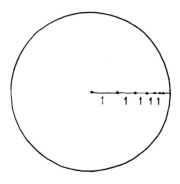

So how do you know that we don't live inside something like this? The further we go, the less we can know what happens on the other side—or even if it means anything to speak of "another side".

But we can still raise the question: what kind of universe do we live in? Then the mathematician creates models, and the physicists figure out which of those models fit the world we live in. It is not clear what we mean by "other side". By conceiving the hyperbolic plane in another way, not embedded in the ordinary plane, but intrinsically, by itself, there is no "other side". One of the possible questions is whether our universe is embedded in another one. But then we could not have any direct contact with this other universe, and we would have to deduce its properties only by its effect on our own universe.

All right, let's go back to mathematics. I have this model, and I can make identifications, just like in the euclidean model.

[*A hand goes up.*]

SERGE LANG. Yes?

GENTLEMAN. You defined the distance with respect to the center, but can one define the distance also for any two points?

SERGE LANG. Yes, of course, but it would be much more technical to do so, and the formula would be more complicated, so I don't want to do it. I would need hyperbolic functions to write it down.

OK, so I want to make identifications. I need certain types of motions, which preserve the hyperbolic distance. As before, I can define translations. Suppose I am given a hyperbolic line. It gives me a direction, and I have the notion of hyperbolic distance. Take any point P. Where are its translations in the direction of this line? Along which curve does the point P move, in the direction of the line?

AUDIENCE. ???

SERGE LANG. Well, let A and B be the two end points of the hyperbolic line, as shown here on the figure. The translates of P in the direction of the line are going to be on the arc APB.

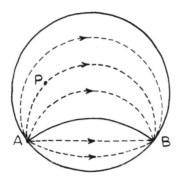

I can translate in one direction, or in the reverse direction, or iterate translations, and so on. Translations are examples of motions which preserve distance. There are others. Do you known which?

A HIGH SCHOOL STUDENT. Rotations, reflections.

SERGE LANG. Exactly. And in the hyperbolic plane, rotations are the same as in the Euclidean plane. As for reflections I am going to draw a point P and its reflection with respect to a line.

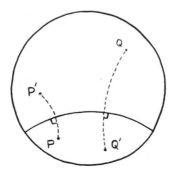

I can also draw a triangle, and its reflection with respect to the same line.

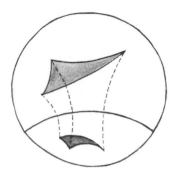

The reflection of a point with respect to the center is the same as in the euclidean case.

LADY. Then the hyperbolic plane has a center?

SERGE LANG. No, the hyperbolic plane by itself does not have a center, but the model which I gave for it has one. In the hyperbolic plane, the situation looks the same everywhere, no matter how close you are to the boundary for the euclidean distance. Given any two points P and Q there always exists a translation which moves P and Q. One says that the hyperbolic plane is homogeneous. In the hyperbolic plane, you are always infinitely far from the boundary. When I represent it by a disc, I choose a center, just as when I choose an origin for the euclidean plane.

OK, let's go back to identifications. I have rotations, translations, and reflections, and I can also combine them with each other, iterate them. In general, these give me all distance-preserving transformations. As I already said, I shall call them motions for short.

Now I still have to define another notion, that of a group of motions. I shall say that:

Γ is a group of motions if:
  1) when two motions M_1 and M_2 are in Γ, then their composite $M_1 M_2$ is also in Γ.
  2) the inverse of a motion M in Γ is also in Γ.

The composite $M_1 M_2$ is the motion such that, when you apply it to a point P, you get $M_1(M_2(P))$. The inverse of a motion which sends P to Q is the motion which sends Q to P. So now we have the notion of a group of motions.

I also need the notion of a discrete group. Let's start with an example, in the ordinary plane, and with translation. Take a point P, and translate it.

$$P \qquad \tau(P) \qquad \tau^2(P) \qquad \tau^3(P)$$

A STUDENT. The points are all distinct.

SERGE LANG. Yes, and what happens if I take a bounded domain in the plane?

THE STUDENT. Eventually, the points get out of the domain.

SERGE LANG. That's right. And I can do the same thing with a group. I will say that two points P and Q are equivalent with respect to the group Γ if there exists a motion M in Γ such that

$$Q = M(P).$$

This is an equivalence. And I say that Γ is discrete if, given a point P, among all the possible motions $M(P)$ with M in Γ, there is only a finite number lying in a bounded part of space. Essentially this means that in any bounded part of space, in any bounded domain, there is only a finite number of points which are equivalent to P with respect to Γ.

To distinguish this new equivalence from rubber equivalence, I have to mention Γ explicitly, so I could all it a Γ-equivalence for short.

Now suppose that Γ is discrete. I can identify points like that, with respect to Γ. I can identify all the points which are Γ-equivalent to each other. Then I obtain a new space after having made these identifications. And I denote this space by

$$\Gamma \setminus H^2.$$

This space will again be two-dimensional.

I have just done identification in dimension 2. Of course, I can also do identifications like this in dimension 3. What do we take as a model for hyperbolic space in dimension 3?

GENTLEMAN. The sphere.

SERGE LANG. Yes, the interior of the sphere, the ball. In dimension 3, we have H^3, which is the ordinary ball, but with a hyperbolic distance which is analogous to the hyperbolic distance in the plane. When you move toward the boundary, the distance becomes arbitrarily large.

And in hyperbolic 3-space, what do planes look like?

GENTLEMAN. They are parts of spheres?

SERGE LANG. That's right.

And we can also define translations, reflections, etc.

But there is also another way of constructing three-dimensional spaces, by using something having dimension 1 and something else having dimension 2. I have already used the product construction before. Now who can give me another example of a three-dimensional space which we could use by taking a product?

THE STUDENT. Take a line and a hyperbolic plane.

SERGE LANG. Ah! Very good! That's precisely what I wanted to get out of your head. So we have another example, by taking the product of the hyperbolic plane H^2 and the line R, which we write as

$$H^2 \times R.$$

This space has the hyperbolic distance on H^2 and the ordinary distance on R.

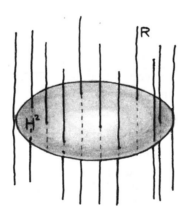

Now we have the fundamental examples

$$H^3 \quad \text{and} \quad H^2 \times R.$$

And we are ready to make the connection with rubber geometry, and the rubber equivalence. I am first going to recall a classical theorem on surfaces. Let's go back to our polygon, which we take in the hyperbolic plane. Its sides are hyperbolic line segments, but the polygon is equivalent to a polygon which we already drew, to construct surfaces by identifying certain sides of the polygon.

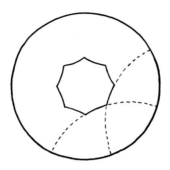

Then the hyperbolic plane can be covered by translations of this polygon, such that two translations are either disjoint, that is, they have no points in common, or they meet only along some side.

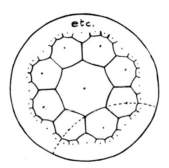

It's the same thing as when you cover the euclidean plane with squares, or rectangles, except that you cannot get a tiling of the euclidean plane by means of regular octagons, but you can tile the hyperbolic plane with any regular polygon.

Identifying certain sides as we did before amounts to making identifications with respect to a group of translations. And one has the theorem:

Theorem. Let F be a surface, compact, orientable, without boundary, and not equivalent to the sphere or to the torus. Then there exists a discrete group Γ such that the surface F is equivalent to the hyperbolic plane on which we have identified points with respect to Γ. In other words,

$$F \sim \Gamma \backslash H^2.$$

Well, this theorem dates back to the 19th century. And nobody, until Thurston, thought that there could be anything like it in dimension 3. It was Thurston's great discovery that there should be an analogous result, to conjecture it, and to prove it in certain cases. First I am going to state a result which connects the first part of the talk on rubber geometry, with the second part on non-euclidean geometry.

We always denote by Γ a discrete group of motions, but we suppose in addition that:

—For any point P, the only motion M in Γ such that $M(P) = P$ must be the identity, that is the motion which does not move any point.

—And to keep me honest, that the motions in Γ preserve orientation.

We always suppose that Γ satisfies these two extra conditions, even if I don't say so explicitly.[8]

Let Γ be a group of motions of H^3, for instance. Then we can have two cases.

First case. $\Gamma \backslash H^3$ is compact.

In this case, the space we get by making identifications with respect to Γ is a manifold of dimension 3, compact, and without boundary. This is one of the ways of obtaining such manifolds.

Second case. $\Gamma \backslash H^3$ is not compact.

In this case, we have to use not only the notion of distance, but also the notion of volume which comes along with it. After I identify with respect to Γ, it is possible that the space $\Gamma \backslash H^3$ has finite volume. I will always suppose that Γ denotes a group such that the volume of $\Gamma \backslash H^3$ is finite.

Of course, you can have the same phenomenon in dimension 2. You can have a polygon whose sides go toward the boundary, and so the polygon has ends which are arbitrarily far from the center, as on the following figure.

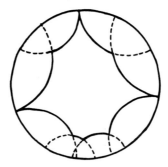

---

[8] This is indeed the case for translations, and these conditions eliminate the possibility of reflections being in Γ.

But you can have a group of motions Γ, and even a group of translations, such that the surface which you obtain by making identifications with respect to Γ has ends which tend to infinity.

The same thing can happen in dimension 3 but it's too hard to draw. The parts which tend to infinity are sort of tubes.

Instead of H^3, I could also take $H^2 \times R$, and consider groups Γ such that $\Gamma \backslash (H^2 \times R)$ is compact, or is non-compact but has finite volume, with tubes going to infinity. What do these tubes look like?

Theorem. Let Γ be as above, a discrete group of motions of H^3 or of $H^2 \times R$, and let's make the identifications with respect to Γ. Suppose that the space you get has finite volume. Then this space is either compact, or is an octopus. Furthermore, $\Gamma \backslash H^3$ is an irreducible octopus.

In this theorem, it is understood that my group Γ satisfies the extra hypotheses I have stated above, for instance the volume of $\Gamma \backslash H^3$ or $\Gamma \backslash (H^2 \times R)$ is finite.

A LADY. In addition to the fact that Γ is discrete?

SERGE LANG. Yes, in addition, it's an extra hypothesis I have to make. After the identifications, I have to suppose that the volume is finite.

There is also a converse, which already gives some idea about the classification of octopusses.

Theorem. Every irreducible octopus is equivalent to a space of type

$$\Gamma \backslash H^3 \qquad \text{or} \qquad \Gamma \backslash (H^2 \times R).$$

And so I get back the octopusses! It's quite extraordinary. We started from an entirely different point of view, we made identifications in a geometry with distances, we took motions preserving the distance, and what do we find? Compact manifolds and octopusses! This is the connection between the first part with rubber geometry, and the second part with geometry of distances.

We are now coming to Thurston's conjecture. We have just seen two examples, H^3 and $H^2 \times R$, which are spaces with distances. I talked about a well-defined and short list of spaces. It consists of:

$$R^3, \qquad S^3, \qquad S^2 \times R, \qquad H^3, \qquad H^2 \times R,$$

that's five, and three others which I don't write down because it would be too technical to do so. There are eight of them altogether. Let's denote by X any one of these spaces.

Then Thurston's conjecture can be stated like this.

Conjecture. Let V be a three-dimensional manifold, compact, without boundary, and always orientable so we don't make things too complicated. Suppose that V is irreducible for the sum along spheres. Then V is equivalent to one of the following cases.

—There exists a unique X among the eight, and a group Γ such that X is compact and $V \sim \Gamma \backslash X$.

—V is a finite sum of octopusses, and each octopus is equivalent to some $\Gamma \backslash X$, where $X = H^3$ or $X = H^2 \times R$.

Besides, in the second case, there is a sort of uniqueness. More or less, this means that if we write V as a minimal sum of octopusses, then the expressions $\Gamma \backslash X$ which occur in this sum are essentially uniquely determined, up to an appropriate equivalence. It would be too technical to make this precise, and to define exactly what we mean by "essentially". One would have to define new equivalences, and this is not the time to do it.

So you see, to get octopusses, all you need is H^3 or $H^2 \times R$. This is the theorem which Thurston's is trying to prove, and which he has proved to a large extent.[9]

GENTLEMAN. What about Poincaré's conjecture?

SERGE LANG. S^3 is in the list, and Γ in this case is the identity. Poincaré's conjecture is isolated at one end of the list, and there is nothing you can do about it.

LADY. I have lost sight of something. What is the difference between R^3 and S^3?

SERGE LANG. R^3 is not compact, it's the ordinary euclidean space around us. But S^3 is like the sphere, it is compact, while R^3 goes to infinity.

[*A hand goes up.*]

SERGE LANG. Yes.

LADY. Can you recall the definition of "discrete"?

SERGE LANG. "Discrete" means that if P is any point and we move P with all possible motions in Γ, that is you look at all points $M(P)$ with M in Γ, then in any bounded part of space there is only a finite number of such points.

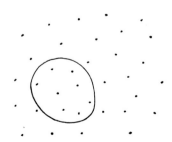

---

[9] W.P. Thurston, "Three dimensional manifolds, Kleinian groups, and hyperbolic geometry," *Bull. Amer. Math. Soc.* Vol. **6**, 3, (1982).

That's what it means, that Γ is discrete. Plus the additional hypotheses I made.

A HIGH SCHOOL STUDENT. And if you take a group which is not discrete, what do you get?

SERGE LANG. Something disgusting, it's lousy. [*Laughter.*] No, it's more complicated, it's not a surface. If the group is not discrete, first you must ask that it be closed to get something which is half way decent. But if it's closed, then the dimension is going to go down. However, if the group is discrete, then there is lots of space between any two points which you identify, and the dimension stays the same when you identify points. If the group is not discrete, then it can give you something horrible—well, not necessarily horrible, but the dimension goes down. OK, you can look up what happens in books, to find out what goes on.

GENTLEMAN. A while back, you defined manifolds by using a hyperbolic geometry, that is the notion of hyperbolic distance. Was this just to fix ideas, or does Thurston's theory depend on the notion of neighborhood, open and closed sets, which are more general notions that that of hyperbolic distance?

SERGE LANG. When you speak of neighborhoods, you are dealing precisely with a type of geometry for which there are no distances. I made a list of eight geometries:

R^3, S^3, and $S^2 \times R$ with the ordinary distance;
$H^2 \times R$ with the hyperbolic distance on H^2 and the ordinary distance on R;
H^3 with the hyperbolic distance;

and three other cases, which have more complicated distances, which can't be called hyperbolic.[10]

---

[10] For mathematicians, I include here the description of these three geometries. One of them is $\widetilde{PSL_2(R)}$, where the $\sim$ denotes the universal covering space. The last two are groups of matrices. One of them consists of the matrices

$$\begin{bmatrix} 1 & a & b \\ 0 & 1 & c \\ 0 & 0 & 1 \end{bmatrix}, \text{ which is the Heisenberg group.}$$

The other consists of the matrices

$$\begin{bmatrix} a & 1 & b \\ 0 & a^{-1} & c \\ 0 & 0 & 1 \end{bmatrix}, \text{ with } a \text{ real and positive.}$$

The underlying spaces of these two are equivalent to R^3, but the distances are different from the usual euclidean distance.

THE GENTLEMAN. Then the topological properties of octopusses are closely tied up with these distances, whereas a priori they should be independent of them?

SERGE LANG. This is an excellent remark. Thurston's discovery was precisely that there was a connection between the two types of notions. This is what gives a lot of kick to his theory.

THE HIGH SCHOOL STUDENT. Yes, but you should then classify the discrete groups of the hyperbolic plane, otherwise, it won't work.

SERGE LANG. [*Laughing and very pleased*] He is absolutely right. What's your name?

THE STUDENT. Paul.

SERGE LANG. Indeed, what have we done here? We reduced something we didn't know to something else which we know—or which we don't know.

So, in the history of mathematics, it turns out that in certain cases, we know discrete groups pretty well, but not in other cases. One knows a lot of things about them, but many remain quite mysterious. But many people have worked on them, in the nineteenth century and the twentieth century. During the last thirty years, there has been a lot of progress about these groups. We know some of them very well. Similarly, one knows some three-dimensional manifolds quite well, and others not at all. So my answer is that, by reducing the study of manifolds to quotients of geometries with distances by means of discrete groups, one had the impression of making a huge step forward. My answer is therefore relative.

You see, in mathematics, it can happen that there are two things we don't know anything about, and we prove that one is equivalent to the other. This does not mean no progress has been made. The problems have been cut by half. [*Laughter.*] But this is not quite what happened here. One knew about three-dimensional manifolds in a certain way. One knew about discrete groups another way. In some sense, these ways were complementary. By putting them together, Thurston contributed to understanding both of them.

This does not mean that I personally know the classification of discrete groups. It's not my side of mathematics. I could learn it, but I do something else. I know some examples, and I could give you several if you want, but I don't know them well for the most part. There is no point getting a hang-up about not knowing them. There are lots of things in mathematics. When one needs something, one can always ask a friend to explain it, just the way I asked Walter Neumann to explain Thurston's theory to me.

GENTLEMAN. Let's go back to one less dimension. By identifying a square, you get a torus. But what about the sphere?

SERGE LANG. S^2 occurs on the side, you can't get it by identifying something, at any rate not the way I have described it here. Similarly, S^3 is on the side. That's Poincaré's conjecture, to prove that a three-dimensional manifold, which is compact, without boundary, and without holes, is equivalent to S^3. This conjecture remains all alone on one side of the theory. The difficulties which come up in connection with S^3 are different from those in connection with octopusses. Poincaré's conjecture is irreducible.

GENTLEMAN. And in lower dimension, for S^2?

SERGE LANG. For S^2, no problem, one knows the answer since the 19th century, that a surface of dimension 2, without holes, without boundary, and orientable, is equivalent to S^2.

GENTLEMAN. And one can get it from a representation of the plane . . .

SERGE LANG. No . . . euh . . . what kind of representation?

GENTLEMAN. With the disc H^2.

SERGE LANG. And with discrete groups? No. There is a theorem which says no. If you take the disc, with the Poincaré–Lobatchevski geometry, and you take a discrete group of motions, and you make identifications, then you will never find something equivalent to S^2. This is a theorem. Are you a mathematician?

GENTLEMAN. No.

SERGE LANG. Anyway, it's clear. A mathematician would have known the answer. [*Laughter.*] Oh no, no! No kidding, the question was very relevant, it is quite remarkable how well you react.

LADY. But Poincaré described two such geometries, it seems to me.

SERGE LANG. Well, we are coming back to the gentleman's question a while ago. He said one could put many distances on the same space. There is not only the distance I mentioned, in the hyperbolic plane, when the rate of change of the distance is $1/(1 - r^2)$. There are many other ways of defining distances. There is an infinite number of such ways. The study of such distances is called differential geometry. It consists in studying all possible ways of defining distances, and of introducing certain equivalences and classifying the distances up to such equivalences. But to do this would require an entire course in differential geometry. You are right, the subject is wide open in many directions.

LADY. But concretely, there is no realization . . .

SERGE LANG. Ah, concretely. But what one person finds concrete, another person will find abstract. It's entirely relative to your own brain, to what you know, to your talent in mathematics, to your intelligence, to your tastes, to your feelings. It's entirely relative. There is no absolute notion of what is concrete and what is abstract. For instance, what you might have found too abstract yesterday or today, could become concrete for you tomorrow.

If I draw enough octopusses, they will appear very concrete to you, it's a question of habit, partly. It depends on circumstances. There is no absolute answer. Of course, a mathematician could do something which others don't understand. The others could have the psychological reaction to find that it's too abstract, and they would say that instead of saying: "I don't understand."

LADY. It has no reality.

SERGE LANG. "Reality" where?

LADY. Physical.

SERGE LANG. Oh! The world of physics is much more extensive than you think. First, you know that if you take the three spatial dimensions, plus the time dimension, you already get four dimensions. And if you go very far, what do you find? Do you find octopusses? You find four-dimensional things? It could already have physical reality. Where do you stop with your physical reality? In what kind of space do we live? Is it curved? Is it an octopus? Is it something like H^3, or the ball with another metric? It's the physicist's business to find which space, and what kind of metric. It's for the physicist to choose between different models, which have been discovered by mathematicians, or to construct new ones which might fit better. Usually, people think that our space is homogeneous. Maybe this is not the case.

Take a point which wanders in space. Besides its spatial coordinates, there is a time coordinate, but there is also the speed, acceleration, curvature, which give me other parameters, other numbers, other dimensions. Take an electron which moves in space. At the same time, it turns, it wiggles, that gives me other dimensions. It's complicated to give a model for the electron, or even to know if the notion of electron makes sense. To describe those things that wiggle, elementary particles, you need other models, which may come precisely from differential geometry, among other things. Physics doesn't stop in any particular place! It's not just the physics of the drawings I can do here on the blackboard. And for other physical phenomena, maybe I need other models, which will appear too abstract for you today.

LADY. Yes, of course [*and the lady makes a gesture which shows that she has understood that those mathematical models which can be used in physics may come from any theory, no matter how abstract or advanced it may be*].

SERGE LANG. So, a good physicist is somebody who won't be scared by complicated models, who won't be chicken, who will seek his models in what engineers find too abstract. Except that the physicist will find a good model, and he will win. He will make it in the history of science precisely because he will liberate himself from the intellectual constraints of his colleagues, and will make concrete what others found too abstract. In other words, there are no limits. The only limits are for each individual, those of his own brain, his own temperament, his own tastes . . .

[*Serge Lang stops here and catches his breath.*]

Ouf! [*Laughing*] Some marathon!

[*Warm applause. After three and a half hours, there remain about 100 persons in the room.*]

Well, so this is goodbye. It doesn't happen every day, it's unique, to have been able to stay here like this for three and a half hours, with an audience like you. It's unique. I really appreciate it a lot. I was really pleased.